SpringerBriefs on Pioneers in Science and Practice

Volume 10

Series Editor

Hans Günter Brauch

For further volumes:
http://www.springer.com/series/10970
http://www.afes-press-books.de/html/SpringerBriefs_PSP.htm

Lourdes Arizpe

Lourdes Arizpe

A Mexican Pioneer in Anthropology

Organización
de las Naciones Unidas
para la Educación,
la Ciencia y la Cultura

Cátedra UNESCO de Investigación
sobre Patrimonio Cultural Intangible
y Diversidad Cultural
Universidad Nacional Autónoma de México

Springer

Lourdes Arizpe
Centro Regional de Investigaciones
Multidisciplinarias
National Autonomous University
of Mexico
Cuernavaca, Morelos
Mexico

The cover photo was taken by Edith Pérez Flores during the presentation of the Award for Academic Merit of the University of Veracruz, Mexico to Lourdes Arizpe in 2010 and is reproduced with the permission of the photographer.

ISSN 2194-3125 ISSN 2194-3133 (electronic)
ISBN 978-3-319-01895-9 ISBN 978-3-319-01896-6 (eBook)
DOI 10.1007/978-3-319-01896-6
Springer Cham Heidelberg New York Dordrecht London

Library of Congress Control Number: 2013946882

Printed on acid-free paper

Springer is part of Springer Science+Business Media (www.springer.com)

To all the people who stand firm, even under difficult conditions, to ensure that human reason and effort will lead us to a conceivable future

Flowering for the Day of the Dead in Amilcingo, Morelos, Mexico, in 2004. *Source* The photo was taken by Edith Pérez Flores and is here reproduced with the permission of the photographer.

Books by the Same Author Published by Springer

- Lourdes Arizpe, Cristina Amescua: *Anthropological Perspectives on Intangible Cultural Heritage.* Springer Briefs in Environment, Security, Development and Peace, vol. 6 (Cham–Heidelberg–Dordrecht–London–New York: Springer-Verlag, 2013).
- Lourdes Arizpe: *Migration, Women and Social Development: Key Issues.* Springer Briefs on Pioneers in Science and Practice No. 11. Subseries Texts and Protocols No. 5 (Cham–Heidelberg–New York–Dordrecht–London: Springer-Verlag, 2014).
- Lourdes Arizpe: *Culture, Diversity and Heritage: Major Studies.* Springer Briefs on Pioneers in Science and Practice No. 11. Subseries Texts and Protocols No. 6 (Cham–Heidelberg–New York–Dordrecht–London: Springer-Verlag, 2014).
- Lourdes Arizpe: *Beyond Culture: Conviviability and the Sustainable Transition.* SpringerBriefs in Environment, Security, Development and Peace, vol. 13 (Cham–Heidelberg–Dordrecht–London–New York: Springer-Verlag, 2014).

Preface

As a scholar, Lourdes Arizpe has the rare quality of having achieved distinction in anthropological research, science leadership, and international policy-making. This is because her writing and her advocacy have deep roots that interconnect languages, cultures, and countries. Early in her career, she pioneered anthropological research in new fields while being fully committed to advocacy for indigenous peoples and for women. She carried out ground-breaking research on women migrants and women workers that helped consolidate women's studies as a field for research. As international scientific organizations increasingly invited her to lead international projects and programs, her research took on a global vision as she headed projects on development and global environmental change. While holding decision-making positions in Commissions and agencies of the United Nations, she brought her analytical skills and her broad experience with development and environmental projects to bear on international policy on culture and on human sustainable development.

There is a third aspect of Lourdes Arizpe's work that is not often recognized. Her deep empathy and her ability to quickly understand the intellectual and political dilemmas at stake in discussions allow her to bring people together in finding common ground to move forward. Once, at the end of one of her conferences at an American Anthropological Association meeting, I overheard one colleague refer to her leadership, and the person he was talking to replied, "Actually, she doesn't lead. She inspires." When she made a huge success of the 1993 World Congress of the International Union of Anthropological and Ethnological Sciences in Mexico City, attended by more than 8,000 anthropologists, we all agreed that her leadership style inspires. She had invited all Mexican students of anthropology to be hosts and guides for foreign students and anthropologists, thus laying the ground for research networking for the future.

Knowledge and Ethics

Lourdes Arizpe's work is driven by an unrelenting quest for knowledge which is analytical and rigorous, yet committed to the highest ethical purpose. The multiple cultures in her home gave Lourdes Arizpe a worldview from her earliest experiences. While her father was Mexican and her mother Swiss, for her primary and secondary education she attended an American school in Mexico City and then went on to study to become an interpreter at the University of Geneva. Soon after, she received an M.A. from the National School of Anthropology in Mexico, and a Ph.D. from the London School of Economic and Political Science.

In her early research, influenced by Imre Lakatosh, with whom she had studied at the London School of Economics,[1] she set out to try and find the 'first principles' in a research program covering the social and cultural dimensions of society. Soon after, under the mentorship of Rodolfo Stavenhagen at the Colegio de México, she participated in an important movement in Mexican and Latin American social science linking research and political advocacy.

In her first, in-depth ethnography of a monolingual Nahuatl-speaking village in the Sierra de Puebla,[2] her analysis showed that family composition had shifted from extended and compound families to nuclear families as the ancient maize production system gave way to a cash crop, coffee.[3] Her second major study on Indian migration opened a new field of research on the selectivity of migrants as she discovered that ethnicity and gender created differentiated patterns of migration within the rapidly increasing rural-urban flows to Mexico City. Importantly, her analysis of fieldwork data demonstrated that Indians generally and women in particular were agents who made deliberate choices in contexts of poverty and marginalization that helped them stabilize and improve their situations. Accordingly, in her first study of the 'Marías', Mazahua[4] women street vendors in Mexico City, she developed a model that helped explain the selectivity of migrants according to individual motivation, middle-level causes and macro-level causes as push factors in migration.[5] She also showed that selling fruit seasonally in Mexico City was the best choice they could make in difficult conditions of marginalization and poverty.

As she delved deeper into the influence of the domestic cycle of families on their migration patterns, she discovered that rural families facing an increasing economic deficit[6] developed a strategy of 'relay migration'.[7] The step-wise pattern

---

1 She was also influenced by her teachers Jean La Fontaine, I. M. Lewis, Maurice Bloch and Julian Pitt-Rivers, her tutor at LSE.

2 Arizpe (1972).

3 As the first chapter of this book shows, she traced this shift to the kinship terminology system in which the Spanish term for cousin, *primo*, was then added to the Nahuat term for brother, *icniuh*, since, in contrast to extended families, in a nuclear family residence creates a distinction between brothers and cousins: *icniuh* and *primo-icniuh*.

4 The Mazahua, an indigenous group, live in a region to the Northwest of Mexico City.

5 Arizpe (1975). A shortened version was published in English: Arizpe (1977).

6 In the 1950s, the Mexican government's economic policy of subsidizing the growth of urban industry kept the price of maize, the staple food, from rising, and fostered an increasing economic deficit in households which families then tried to offset by sending their children off to the city to send back cash remittances.

7 Arizpe (1980).

she found showed that families sent out their children, especially young girls, who could always find work in paid domestic labour, to wage labor in the city so they could send back cash remittances needed to continue to cultivate maize. Her study also showed that villages have collective arrangements for reciprocity and mutual aid in the continuous transit of migrants to and from the village and the city. Based on her observations of the impact of development policies and of demographic growth in these rural areas, she predicted that rural–urban migration would increase exponentially in Mexico, as indeed it has, as in most developing countries.

At the same time as she was conducting research, Lourdes Arizpe participated actively in the initial thrust of the rising tide of indigenous movements in Mexico and Latin America. She wrote extensively on this in Mexican newspapers and journals during the 1970s and 1980s. Two of her articles have been included in this book, because they illustrate the context in which discussions on ethnicity and gender were starting to be held at that time. They describe an important event, the First National Congress of Indians held in Pátzcuaro, Mexico in 1975; they were subsequently published by the National Indianist Institute in *The Challenge of Cultural Pluralism.*

By the end of the 1970s, her research had become widely known and disseminated in other developing countries, and she accepted invitations to spend prolonged periods of time, in some cases including fieldwork, in India, Bangladesh, and Senegal through a John S. Guggenheim fellowship.[8] She also received a Fulbright scholarship to teach at Rutgers University in the United States. Soon after, she was invited to participate in the international group that organized the North–South Dialog and in the Society for International Development. Subsequently, she continued to collaborate closely with the group that launched the Human Development Report at the United Nations Development Program[9] and, more recently, the capabilities approach in Human Development.

Ethnicity and Development

During the 1970s, as she had developed her work on ethnicity, Arizpe asked whether Indians were poor because they were Indians, or whether they were Indians because they were poor. She surprised everyone by arguing that both statements were true: Indians were poorer because they were discriminated against and this excluded them from development opportunities. Yet she had also found, in her long conversations with them, that they themselves, in many cases, chose to

[8] She was sponsored for such work by the National Indianist Institute of Mexico (1978), a Fulbright scholarship (1979), a John S. Guggenheim fellowship (1981), and the *International Labour Organization* (1983–1988) and UNESCO (1985–1987).

[9] She collaborated in several projects with Amartya Sen, Mahbub ul Haq, Sakiko Fukuda-Parr, Francis Stewart, Louis Emmerj and Paul Streeten, among others.

retain their cultures in spite of the high price they were paying for this, knowing that they would have greater difficulties in having access to markets and to information. Having developed this argument further, 20 years later, at the 25th anniversary of the OECD in 1989, Arizpe surprised the ambassadors present by arguing that sustainability could indeed not be achieved without taking into account cultural and social sustainability, which meant allowing different peoples to manage their traditional ways of life as they chose for development. "In many countries," she argued, "some groups, consciously and willingly, will want to stay out of a modernizing development, for the time-cycles in which they live are circular and, in fact, timeless. And their wishes must be respected…". Her article "On the Social and Cultural Sustainability of World Development",[10] published in the journal *Development* in 1989, was re-edited in the same journal in 1995 and was widely disseminated.

Importantly, in that seminal paper, she explained that this group, "…who don't want to change…. must be distinguished from those who are retrenching back into traditional customs or religions because they are being marginalized". Yet a third group, she pointed out, use cultural emblems for seeking comparative advantages over other groups in a given setting. This choice, she had insisted since 1978,[11] should be called 'ethnicism' as it made use of cultural emblems to seek political advantage, setting the term apart from 'ethnicity' which is the intrinsic capacity of any group to create distinctive cultural expression in a given cultural group. She ended by proposing that legislation should be changed, especially in countries of the South, "to ensure a balanced political participation of all ethnic groups." Hence, she had pulled together questions and thoughts that were being discussed in both academic and NGO forums into a coherent set of notions on cultural pluralism which just a few years later, in the UN *World Commission on Culture and Development* (WCCD) in 1992–1995, would become central to its report *Our Creative Diversity*.

Women and Development

After her pioneering study on the 'Marías', Lourdes Arizpe extended her study to analyze the situation not only of Indian women but also of all women in the context of development. Her study of women workers in the strawberry agribusiness in Zamora, Mexico was also ground-breaking in that it showed that young girls from peasant families actually preferred to work in the strawberry-packing plants because it gave them freedom from the traditional confinement in unpaid domestic work in their homes. As in her previous research, Arizpe placed their situation in the broader context of development, as shown by the title of an article

---

[10] Arizpe (1989a).

[11] Arizpe (1978).

she wrote with J. Aranda: "The Comparative Advantages of Women's Disadvantages."[12] This article was widely read and quoted, and republished by Oxford University Press and in several countries.

She had followed the work of Ester Boserup and collaborated with other economists such as Carmen Diana Deere[13] and Lourdes Benería in demonstrating that women were indeed involved in development, but in many instances in situations that added a third workload to their already overburdened domestic, family and social daily life. Arizpe's anthropological research gave a close-up look at women's lives and the choices they made. Many of her articles have been republished in developing countries around the world. She also took the initiative in organizing research seminars on women and development, which included a workshop on methodologies to study rural women in Nicaragua, soon after the Sandinistas came to power.

Her work then led her to participate in fostering policy programmes for rural women. In 1983, as advisor to the Ministry of Agrarian Reform in Mexico, Lourdes Arizpe helped create the first ever *Programa Nacional para la participacion de la mujer en la consecucion del desarrollo rural* (PROMUDER)[14] which placed campesino women for the first time in the mainstream policy arena. She has also been an Advisor for other government programmes on Indian women and on women migrants and in international agencies such as ILO and UNESCO.

In her interviews with Mexican women migrants in the United States Arizpe found that many of them said, quite openly, that they preferred to live in the USA because there they felt that the police protect them against domestic violence. In recent years, she has argued that women are still 'moving boundaries' as the 'caring economy' not only pulls more women toward rich countries but also creates difficult social conditions for their families in their home countries.

Lourdes Arizpe became a feminist after witnessing, time after time, the vulnerability and the total defencelessness of Indian women, in particular, but also, generally, of women in Mexico. It is of no coincidence that a specific crime has been juridically defined and incorporated into law in Mexico: that of 'feminicidio', which is the assassination of women by men who are their husbands or next of kin. In the first decade of this century, feminicides in Mexico have increased markedly in an atmosphere of unprecedented growth of inequality and of organized crime.

At the same time as she was carrying out field studies, Lourdes Arizpe was actively involved in setting up Women's Studies programmes in Mexico.[15] In 1977 she proposed a *Central American and Latin American Congress on Women's Studies*, which about 50 women were expected to attend since women's studies

---

12 Arizpe/Aranda (1981).

13 Arizpe (1987).

14 National Programme for Women's Participation in Achieving Rural Development.

15 She was one of the founders of the feminist journal *Fem* in 1977. She was also one of the founders of the *Programa de Investigación y Estudio sobre la Mujer* (PIEM) at El Colegio de México. In the National University, she also helped to create the *Programa Universitario de Estudios de Género* (PUEG).

were just beginning to become known in the region. In fact, an avalanche of 400 women researchers flowed into to the Congress, including from the United States and other developed countries. Lourdes Arizpe wanted women's voices to be heard and their ideas acknowledged. One of her greatest satisfactions has been to see the growth of women's leadership in all spheres of life in Mexico, as well as in all other regions of the world. Internationally, Arizpe was also very actively involved in setting up international women's NGOs, among them *Development for Women in a New Era* (DAWN) as well as gender programmes in international institutions such as ILO and the *United Nations Institute for Social Research* (UNRISD) where she was Chair of the Board (2007–2012).

The Human Dimension of Global Environmental Change

In her leadership of scientific organizations[16] she encouraged the development of fields of research into the social and cultural dimensions of global environmental change. As always, she insisted on first going into the field to acquire a closer understanding of the human issues side of emerging concerns over the depletion of natural resources and the loss of biodiversity. In her three-year fieldwork study of settler communities in the Lacandon rainforest, in which I myself participated, she developed useful heuristic tools for the social perception of deforestation based on fieldwork carried out with ten young scholars: anthropologists, ecologists, psychologists, and a veterinary surgeon, all from the *National University of Mexico* (UNAM). The research findings were published in Spanish by the National University of Mexico, as well as by the University of Michigan at Ann Arbor in the United States.[17]

During this fieldwork with the team, a number of incidents made it evident that a major conflict was brewing in the depths of the rainforest which no one could yet publicly speak about. On 1 January 1994, three months after our research was published, the Zapatistas came out of the rainforest and declared war on the Mexican state. Lourdes Arizpe was one of the few social scientists who had, at that time, a profound understanding of the environmental and political tensions in the Lacandon region. Given the sweeping idealization of the Zapatistas which followed their appearance in the mass media, Arizpe maintained a critical, independent position, insisting that the Zapatista uprising represented a legitimate

[16] L. Arizpe was first President of the *Colegio de Etnólogos y Antropólogos*, 1985–1988 and Secretary of the Mexican Science Academy, 1992–1994, both in Mexico. She was then elected President of the *International Union of Anthropological and Ethnological Sciences*, 1988–1993. She also held the office of Vice-President 1992–1994, 1998–2002 and then President of the *International Social Science Council*, 2002–2006.

[17] Arizpe/Paz/Velazquez (1993); in English: Arizpe/Paz/Velazquez (1996a).

movement for social justice but that their strategy would not bring improvement to the lives of poor settlers in the rainforest.[18]

Her article on "The Global Cube"[19] published by UNESCO was again an attempt to theorize about the sociosphere in the context of environmental change.

In 1993, Lourdes Arizpe surprised anthropologists by making the cultural dimensions of global change[20] the major theme of the World Congress of the International Union of Anthropological and Ethnological Sciences. A few years later, as President of the *International Social Science Council* (ISSC), she helped focus this question in international scientific research policy by asking the then Executive Director of the *International Council of Scientific Unions* (ICSU; now the *International Council for Science*), in a joint General Assembly of ICSU and ISSC, what percentage of global environmental change was anthropogenic. His answer, "seventy percent," made a deep impression on the natural scientists present. It helped set aside the protracted debates on the primacy of either environmental or social factors as drivers of ecosystemic change. It also gave a greater thrust to the ISSC *Human Dimensions of Global Environmental Change Programme* (HDGEC) which is now a major joint ISSC–ICSU research programme located in Bonn, Germany.

With her in-depth knowledge of global change, as Assistant Director-General for Culture at UNESCO from 1994 to 1998, Arizpe brought to policy-making discussions the question of the deep overlapping of biodiversity and cultural diversity. For the first time, a table was included in the 1998 World Culture Report—which she helped launch—which showed the close correlation between the number of linguistically diverse groups in a country and the biodiversity in their ecosystems.

Influencing the Business Ecosystem

At the Davos Global Economic Forum, Lourdes Arizpe, as a member of the Academic Faculty (2002–2005), was asked to coordinate sessions that would help world business leaders understand the emerging concerns over identity, cultural pluralism, cultural and art policy, and social issues. At the Davos meeting in 2004, as a speaker at the traditional breakfast held by a large food multinational, she warned that her fieldwork data were showing that in rural villages and among groups living in poverty, children were going directly from malnutrition to obesity. Recently, Lourdes Arizpe was invited to participate as a consultant in projects on cultural diversity and development sponsored by business companies. Since 2010, she has been a member of the high level Orientation Committee of the innovative

[18] Arizpe (1996a, 2002).

[19] Arizpe (1991).

[20] Arizpe (1996b).

project 'Ecosysteme Fund' which Danone International has developed in several countries around the world.

Culture, Development and Heritage

In the 1980s, as she continued to try to elucidate the 'first principles' of social life, Lourdes Arizpe extended her fieldwork to non-Indian populations. She made a bold move to analyse culture in a Mexican town by testing Theodor Adorno's survey on the *Authoritarian Personality* and by examining values in a Catholic context.[21] Her findings showed that ethnic and religious beliefs had a lesser impact on people's general ideological outlook than gender and their position within the market economy. Women in the highest social class and women in peasant families were the most conservative in outlook, while, in contrast, men of the highest social class and women agricultural workers were radically open to change.

This convinced her that development policies urgently needed to include culture as an analytical heuristic in making development a more adaptable, endogenous process. In 1992, she was invited to be a member of the independent United Nations World Commission on Culture and Development and, soon after, was placed in charge of its Secretariat for the writing of its report *Our Creative Diversity*.[22]

By then, Lourdes Arizpe had been designated Assistant Director-General for Culture in UNESCO and was awed by the fact that she had to manage over 400 cultural projects in 182 countries around the world. You could hear her saying, often, that she marvelled at the fact that peoples speaking so many different languages from so many different cultures, could work together in such tightly defined, geopolitically sensitive and pressured programmes. She also coined a memorable quote: that "working for the United Nations was like working for a government in which all political parties are in power at the same time".

Lourdes Arizpe was then able to bring all her previous experience in fieldwork and in anthropological analysis to organizing UNESCO's programmes. Highlights of her contribution to UNESCO were projects on indicators of culture and development, global ethics, gender and culture, and the overlapping of cultural, social and eco-biological processes. According to her colleagues at the Culture Sector of this institution, one of her contributions was to have moved UNESCO's culture sector "from archaeology to social anthropology", that is, to have brought a dynamic perspective to cultural policy-making and programs. Against huge odds, she was able to get UNESCO, for the first time, to publish two World Culture Reports that included new statistical and thematic research on culture.

She was able to bring her experience as a Researcher and as Director of the National Museum of Popular Cultures of Mexico (1985–1988) to building the

---

21 Arizpe (1989b).

22 United Nations Educational, Scientific and Cultural Organization (1996).

conceptual foundations which led to the 2003 Convention for the Safeguarding of Intangible Cultural Heritage and to contribute to the scientific committee of UNESCO's World Report on *Investing in Cultural Diversity and Intercultural Dialogue*.[23] In Mexico, in the past decade, Lourdes Arizpe has led research projects on culture in which more than 50 researchers have been involved. At the Center for Multidisciplinary Studies of the National University of Mexico, she is now coordinating a UNESCO UNITWIN Chair on Intangible Cultural Heritage and Cultural Diversity and also chairing the International Social Science Council's Commission on Intangible Cultural Heritage.

Still UNESCO UNITWIN Chair concerned that multiculturalism is creating a world in which an archipelago of cultures and religions are fostering ever-growing conflicts and reterritorializations through ethnic and religious cleansing, Lourdes Arizpe has endeavored to look for an alternative perspective. At present, she is writing a book on the 'First Principles' that must be put in place to build more convivial social structures in a world overloaded with information and communication.

As she wrote in one of her essays: "*Le pari*[24]: the challenge, is to shape something that is not yet born but which, in the process of being tried out, will be born."

For all these reasons. Lourdes Arizpe has become not only a *Mexican Pioneer in Anthropology* but also a global intellectual leader on indigenous people, an innovator in women's studies, a global scientific leader who has inspired both the international research and policy communities. She received an Honorary Doctorate from the University of Florida at Gainesville in 2010. She helped place the issues of the impact of ethnicity and gender in migration, the status of women in development, social and cultural sustainability, the human dimensions of global environmental change and intangible cultural heritage on the world agenda for knowledge and policy.

I am delighted that on the occasion of Lourdes Arizpe's 70th birthday a selection of her original and innovative work during the past four decades is again available to scholars and especially young researchers and women scholars both electronically and in print. She would want to inspire them to change the world through scientific research, to encourage present and future generations to become pioneers themselves in peaceful change in order to maintain the cultural and ecological diversity of the world we all live in.

May 2013

Margarita Velazquez Gutierrez
Regional Multidisciplinary Research Centre (CRIM)
National University of Mexico
Cuernavaca
Mexico

[23] United Nations Educational, Scientific and Cultural Organization (2009).

[24] In French, *pari* is a challenge but it also means placing a stake in something you wish for.

Margarita Velazquez Gutierrez At present Director of the *Centre for Multi-disciplinary Studies* (CRIM) of the *National University of Mexico* (UNAM), Dr. Margarita Velazquez Gutierrez received a B.A. in Social Psychology from the Metropolitan Autonomous University in Mexico City and a Ph.D. in Sociology of Development from the Institute for Latin American Studies of the University of London in the United Kingdom. She has conducted research on social policy, development, sustainability, and gender which has been published in six books on topics such as desertification, coastal management and building sustainability with a gender perspective. She has received grants from the National Council of Science and Technology of Mexico; the MacArthur Foundation, the Ford Foundation, the Hewlett Foundation and the British Council. International agencies such as UNIFEM, the Population Fund of the UN, FAO, the *Economic Council for Latin America and the Caribbean* (ECLAC) and the *Inter-American Development Bank* (IADB) have asked her to be a consultant. She also held office in the Mexican government in the Ministries of Agrarian Reform, the National Institute for Women, and the National Council of Human Rights.

References

Arizpe, Lourdes, 1972: "Nahua Domestic Groups: The Development Cycle of Nahua Domestic Groups in Central Mexico", in: *Kung. Magazine of the LSE Anthropology Society* (special issue): 40–46.

Arizpe, Lourdes, 1975: *Indígenas en la Ciudad: el caso de las Marías* (México: Sepsetentas).

Arizpe, Lourdes, 1977: "Women in the Informal Sector: the case of Mexico City", in: *Signs*, 3, 1 (Autumn): 25–37.

Arizpe, Lourdes, 1978: *Migración, etnicismo y cambio económico: un estudio de migrantes campesinos a la ciudad de México* (Mexico City: El Colegio de México).

Arizpe, Lourdes, 1980: "Cultural Change and Ethnicity in Rural Mexico", in: Preston, David (Ed.): *Environment, Society and Rural Change in Latin America* (New York, Toronto: John Wiley and Sons): 123–134.

Arizpe, Lourdes; Aranda, Josefina, 1981: "The Comparative Advantages of Women's Disadvantages: Women Workers in the Strawberry Agroindustry in Agribusiness in México", in: *Signs*, 7, 2 (Winter): 453–473.

Arizpe, Lourdes, 1987: "Mexican Agricultural Development Policy and Its Impact on Rural Women", in: Deere, Carmen Diana; León, Magdalena (eds.): *Rural Women and State Policy*. Series in Political and Economic Development in Latin America (Boulder, Colorado: Westview Press): 67–82.

Arizpe, Lourdes, 1989a: "On the Social and Cultural Sustainability of World Development", in: Emmerij, Louis (Ed.): *One World or Several?* (Paris: OECD Development Centre): 207–219.

Arizpe, Lourdes, 1989b: *Cultura y desarrollo: una etnografía de las creencias de una comunidad mexicana* (Mexico City: El Colegio de México – UNAM – Miguel Ángel Porrúa).

Arizpe, Lourdes, 1991: "The Global Cube: Microsocial Models in a Global Context", in: *International Social Science Journal: Global Environmental Change*, XLIII,130: 599–609.

Arizpe, Lourdes; Paz, María Fernanda; Velazquez, Margarita, 1993: *Cultura y Cambio Global: percepciones sociales sobre la desforestación en la selva lacandona* (Mexico City: CRIM/ UNAM – Miguel Ángel Porrúa).

Arizpe, Lourdes; Paz, María Fernanda; Velazquez, Margarita, 1996a: *Culture and Global Change: Social Perceptions of Deforestation in the Lacandona Rain Forest in Mexico* (Ann Arbor, Michigan: The University of Michigan Press).
Arizpe, Lourdes (Ed.), 1996b: *The Cultural Dimensions of Global Change: an Anthropological Approach* (Paris: UNESCO).
Arizpe, Lourdes, 1996c: "Chiapas: The Basic Problems", in: *Identities*, 3, 1–2: 219–233.
Arizpe, Lourdes, 2002: "Freedom of choice, democracy and the Zapatistas". Paper presented at the London School of Economics and Political Science, London, UK, March.
United Nations Education, Scientific and Cultural Organization, 1996: *Our Creative Diversity. Report of the World Commission on Culture and Development* (Paris: UNESCO).
United Nations Educational, Scientific and Cultural Organization, 2009: *World Report. Investing in Cultural Diversity and Intercultural Dialogue* (Paris: UNESCO).

The "Research Dream Team" in Zacualpan, Morelos, Mexico in 2012. *Source* From the personal photo collection of the author

Acknowledgments

This book is filled with people's lives because they were willing to talk to me about them and so my first recognition goes to all of them. Yet this book would not have been possible without the relentless editorial dedication of Dr. Hans Günter Brauch, the Editor of this book series who also did the copy-editing, to whom I am greatly indebted. In writing these articles I was fortunate to have been able to train many young researchers, among them Margarita Velazquez Gutierrez, Cristina Amescua and Edith Pérez Flores to name only a few of countless others, who made fieldwork an interesting and at times merry endeavour. I would especially like to thank Dr. Serena Eréndira Serrano Oswald for style editing, Mr. Mike Headon (UK) for language editing, Dr. Johanna Schwartz, the Editor with Springer-Verlag and Ms. Agata Oelschlaeger, the Producer with Springer-Verlag in Heidelberg (Germany) and the whole production team at Springer Publishers in Chennai, Tamil Nadu, India. Many thanks also to Norma Angélica Guevara for burrowing in many archives to find texts and to the Publications Department of the *Regional Center for Multidisciplinary Research* (CRIM) for their assistance in the editing of this book.

Lourdes Arizpe Schlosser talking about crops in Amilcingo, Morelos, Mexico, in 2012. *Source* The photo was taken by Edith Pérez Flores and is here reproduced with the permission of the photographer

Contents

Lourdes Arizpe as President of the International Union of Anthropological and Ethnological Sciences in 1992. *Source* From the personal photo collection of the author

Part I
On the Author

With Javier Pérez de Cuéllar, former Secretary-General of the United Nations, celebrating that Lourdes Arizpe received the Palmes Academiques of France in 2007. *Source* From the personal photo collection of the author

Chapter 1
Biographical Information of the Author

1.1 A Brief Biography of Lourdes Arizpe

Lourdes Arizpe was born in Mexico City and grew up travelling with her family to all regions of Mexico, an itinerary that drew her attention to the rich diversity of indigenous cultures. Her home was also multicultural, as her Mexican father and Swiss mother, as well as her parents, were multilingual. While at grammar school at an American school in Mexico City, she plunged into literature, both in English and in French, which she studied with a French teacher who opened up her library to her. A few years later she received a First in English Studies at the British Council and the First Place Award in the final course on Advanced Studies in French language and history at the French Institute for Latin America. After taking the SAT examination, they placed her in the Honors English Class at the University of the Pacific in California and she made it to the Dean's List during her stay there. She also studied to be an interpreter in Geneva University, Switzerland and received a Certificate in French Studies. Upon her return to Mexico, she studied history at the National University of Mexico, then an M.A. in ethnology from the National School

Lourdes Arizpe studying world cultures at the London School of Economics in 1975. *Source* From the personal photo collection of the author

L. Arizpe, *Lourdes Arizpe*, SpringerBriefs on Pioneers in Science and Practice 10,
DOI: 10.1007/978-3-319-01896-6_1, © The Author(s) 2014

of History and Anthropology in 1970 and a PhD. in social anthropology from the London School of Economics and Political Science, UK in 1975.

In the diverse fields of research in which she conducted fieldwork in Mexico, Nicaragua, Peru, India and Bangladesh, Lourdes Arizpe pioneered anthropological studies on migration, gender, rural development and culture and global change. Recognition for her work led to invitations to participate in international research networks and institutions. Professor Arizpe taught at Rutgers University through a Fulbright grant in 1979 and carried out research in India and Bangladesh with a John S. Guggenheim grant in 1981. She was very active in the emerging international networks of researchers on women and gender, and was a founder of several initiatives in publications and both national and international academic programmes in this field. She was an active participant of all the United Nations International Conferences on Women.

At the same time that she conducted research, training many students as she went along, Lourdes Arizpe held office in scientific organizations and institutions. She was academic coordinator of the Sociological Centre at the Colegio de Mexico, then director of the National Museum of Popular Cultures. Elected as President of the National Association of Ethnologists of Mexico she became Secretary to the Mexican Science Academy and later on Director of the Institute of Anthropological Studies at the National University of Mexico. She was also elected to the Executive Committee of the Latin American Studies Association (LASA) and of the Society for International Development based in Rome. In 1988 Professor Arizpe was elected President of the International Union of Anthropological and Ethnological Sciences in Zagreb and successfully organized its World Congress five years later in Mexico City in 1993.

With rural development activists in Chennai, Tamil Nadu, India in 1982. *Source* From the personal photo collection of the author

During her academic career she lectured at many universities in Mexico, the US the UK France, Sweden, and India. She also spent prolonged periods of study and sabbaticals at Cambridge University in the UK, at the Institute of Social Studies in the

Netherlands, the Center for Research and Development in Santander and the University of Barcelona in Spain, and at the Maison des Sciences de l'Homme in Paris, France.

Lourdes Arizpe became a member of the United Nations Commission on Culture and Development, and soon after was invited to hold the office of Assistant Director General for Culture at UNESCO. In 2004 she was elected President of the International Social Science Council and participated as a member of the Academic Faculty of the Global Economic Forum at Davos, Switzerland from 2000 to 2004. At the United Nations Institute for Research on Social Development she was Chair of the Board 2005–2011 and a member of the Committee for Development Policy of the Economic and Social Council. She is a member of the Board of Trustees of the Library of Alexandria in Egypt.

With students at the Library of Alexandria, Egypt in 2009. *Source* From the personal photo collection of the author

Lourdes Arizpe became an Honorary Member of the Royal Anthropological Institute of the UK in 1995, received the Order of 'Palmes Academiques' from France in 2007, became a member of the World Academy of Science and received the Award for Academic Merit of the Universidad Veracruzana in Mexico and an Honorary Doctorate from the University of Florida at Gainesville in 2010. At present she is Chair of the recently created Commission on Intangible Cultural Heritage in the International Union of Anthropological and Ethnological Sciences, and Coordinator of the Unesco Unitwin Chair of Research on Intangible Cultural Heritage and Cultural Diversity of the National University of Mexico.

1.2 The Author's Relevant Activities and Awards

1.2.1 Institutional Affiliations

- Researcher, *El Colegio de México* (1972–1985)
- Academic Coordinator, Department of Sociology, *El Colegio de México* (1975–1977)

- Professor, *Regional Centre for Multidisciplinary Research* (CRIM), *National University of Mexico* (UNAM) (since 1988)
- Director, *Institute of Anthropological Research, National University of Mexico* (UNAM) (1991–1994)
- Assistant Director General for Culture, *UNESCO* (1994–1998).

Addressing UNESCO General Conference in Paris in 2005. *Source* Photo was taken by Ph. Sayah Msadek, Paris Photo Press, and is from the personal photo collection of the author

1.2.2 Scientific and International Leadership Positions

- *President*, Colegio de Etnólogos y Antropólogos Sociales, Mexico (1985–1988)
- *President*, International Union of Anthropological and Ethnological Sciences (1988–1993)
- *Vice-President*, Society for International Development, Rome, Italy (1991–1994)
- *Vice-President*, International Social Science Council, Paris (1992–1994; 1998–2002)
- *President*, thirteenth International Congress of Anthropological and Ethnological Sciences, Mexico (1993)
- *Coordinator*, Secretariat of the World Commission on Culture and Development (1994–1995)
- *Coordinator*, launching of the Report of the World Commission on Culture and Development *Our Creative Diversity* in 22 countries of Europe, North America, Asia and South America (1996–1998)
- *Coordinator*, Organizing Committee of the World Conference of the Statute of the Artist, Paris (1997)
- *Coordinator*, Organizing committee, Intergovernmental Conference on Cultural Policies for Development, Stockholm, Sweden (1998)

- *Coordinator*, Scientific Committee, Unesco World Culture Reports (1996–2001)
- *Coordinator*, Working Group on Public Liberties and Human Rights of the Commission for the Reform of the State, Mexico (2000)
- *Chair*, Working Group on Principles, Citizen's Commission against Discrimination, Mexico (2000)
- *Chair*, Committee on Human Rights and Civil Liberties, Mexican Council for the Reform of the State (2001)
- *President*, International Social Science Council, Paris (2002–2004; 2004–2006)
- *Chair*, Board of the United Nations International Research Institute for Social Development, Geneva (2006–2008; 2008–2012)
- *Coordinator*, UNESCO Unitwin Chair for "Research on Intangible Cultural Heritage and Cultural Diversity" at the National University of Mexico (2011)
- *Coordinator*, Commission on Intangible Cultural Heritage of the International Social Science Council (2011–2013).
- *President*, Technical Committee, National Commission for Intangible Cultural Heritage, Mexico (2011–2012).

With Kofi Annan, former Secretary-General of the United Nations, and Frederico Mayor, Director-General of UNESCO, in Paris in 1997. *Source* From the personal photo collection of the author

1.2.3 Selected Other Professional Activities

- Consultant, International Labour Organization (1979–1981)
- Visiting Fellow, Institute of Development Studies, Sussex University (1980–1982)
- Member, Joint Latin-American Committee, *Social Science Research Council*, United States (1987–1990)
- Director, National Museum of Popular Cultures (1985–1988)

- Research Professor, Regional Center for Multidisciplinary Research (CRIM), National University of Mexico (since 1988)
- Member, Executive Committee *International Program for Overcoming Hunger in the 1990s*, Brown University, United States (1990–1993)
- Member of the National Research Council Committee on Environment (1992–1996)
- Member, Executive Committee *Development Alternatives for Women in a New Era* (DAWN) (1990–1994)
- Member, *Board of Trustees, Consortium for Environmental System of Information* (CIESIN), United States (1993–1996)
- Consultant, Folklife Program, *Smithsonian Institute*, United States (1988)
- Member, Executive Committee, *Latin American Studies Association* (LASA) (1994–1996)
- Advisory Committee, *World Report on Human Development*, UNDP (1997–2004)
- Member, *Board for Sustainable Development, National Research Council*, United States (1996–1998)
- Member of the *Advisory Committee for Environment of the International Council for Science* (ACE-ICSU) (1995–2000)
- Member, *Advisory Group on Social Themes for Latin America and the Caribbean, World Bank* (1997)
- Member of the Jury for *Award in International Scientific Cooperation, American Association for the Advancement of Science* (AAAS) (1998–1999)
- Consultant, *International Convention for the Protection of Intangible Cultural Heritage*, UNESCO (1999–2002)
- Member, *Advisory Council of the Project 'History of the Ideas in United Nations'*, Ralph Bunch Institute for International Studies, City University of New York, United States (since 1999)
- Professor, *Faculty of Political and Social Sciences*, National University of Mexico (since 2000);
- Member, Board of Trustees, *Library of Alexandria*, Egypt (since 2004)
- Visiting Professor, *Maison des Sciences de l'Homme*, Paris (2007)
- Member, *Conseil d'Orientation de l'Ecosysteme Danone*, Paris (since 2010)
- Member, *Committee for Development Policy*, Economic and Social Council, United Nations, September (2006–2012)

1.2.4 Awards and Distinctions

1978 *Medal for distinguished services to indigenous peoples*, National Indianist Institute, Mexico
1978 *Fullbright-Hays Scholarship*, United States
1982 *John Guggenheim Grant*, United States
1986 *Member, Mexican Science Academy*, Mexico
1989 *Member, National System of Researchers*, Mexico

1990 *Fray Bernardino de Sahagún Prize*, National Institute of History and Anthropology, Mexico
1992 *Benigno Aquino Prize* for distinguished services to developing countries, William and Mary College, United States
1992 *Member, United Nations World Commission on Culture and Development*
1993 *Gorjanovic-Krambergeri* Prize awarded by the Croatian Anthropological Society, Zagreb, Croatia
1995 *Medal* for distinguished service in the field of culture, Culture Ministry of Pakistan
1997 *Honorary Member of the Royal Anthropological Institute*, UK
2000 *Guest Speaker, American Anthropological Association* Meeting, Chicago
2001 *Member, Group of Eminent Persons for the Dialogue of Civilizations*
2002 *Member of the Academic Faculty of the Global Economic Forum of Davos*
2002 *Visiting Fellow, Andrés Bello Chair, New York University*
2003 *Laureana Wright Award,* Mexican Society of Geography and Statistics
2004 *Award of the Municipal Government of Zacualpan de Amilpas*, Mexico, for revitalizing local intangible cultural heritage
2004 *The Biennial Lourdes Arizpe Award in Anthropology, Politics and Ecology* created at the *American Anthropological Association*
2008 *Member of the National Order of Academic Palms of France*
2010 *Honorary Doctorate, University of Florida at Gainesville*, United States
2010 *Honorary Member, International Union of Anthropological and Ethnological Sciences*
2010 *Member, World Science Academy*
2011 *Member of the Scientific Evaluation Panel* of the European Council for Scientific Research
2011 *Medal for Academic Merit*, Veracruz University, Mexico
2011 *Award* for services in safeguarding the popular and indigenous Cultures of Morelos, *National Department of Popular and Indigenous Cultures*. Mexico
2012 *Member, Board of the National Council to Prevent Discrimination*, Mexico

Receiving Honorary Doctorate at the University of Florida, Gainesville, United States in 2010. *Source* From the personal photo collection of the author

Chapter 2
The Author's Relevant Publications

2.1 Major Books

1972 *Parentesco y Economía en una Sociedad Nahua* (Mexico City: Instituto Nacional Indigenista), 255 p; republished as: *Parentesco y Economía en una Sociedad Nahua* (Mexico City: Instituto Nacional Indigenista, 1990)

1975 *Indígenas en la Ciudad: el caso de las Marías* (Mexico City: Sepsetentas), 180 p; republished as: *Indígenas en la Ciudad: el caso de las Marías* (Mexico City: SEP Ochentas, 1982)

1978 *Migración, Etnicismo y Cambio Económico: Un estudio de migrantes campesinos a la ciudad de México* (Mexico City: El Colegio de México), 261 p

1978 *El Reto del Pluralismo Cultural* (Mexico City: Instituto Nacional Indigenista), 76 p

1982 (Ed.): *La Mujer y el Desarrollo, Vol. II: La Mujer y la Unidad Doméstica* (Mexico City: SEP—Diana)

1986 *Campesinado y Migración* (Mexico City: Secretaría de Educación Pública), 153 p

1989 *La Mujer en el desarrollo de México y de América Latina* (Mexico City: CRIM-UNAM), 271p

1989 *Cultura y desarrollo: una etnografía de las creencias de una comunidad mexicana* (Mexico City: El Colegio de México—UNAM—Miguel Ángel Porrúa), 286 p (Award "Fray Bernardino de Sahagún", 1990)

1991 With Ludka de Gortari (Eds.): *Repensar la nación: fronteras, etnias y soberanía* (Mexico City: CIESAS), 201 p

1993 (Ed.): *Antropología Breve de México* (Mexico City: Academia de la Investigación Científica), 398 p

1993 With Carlos Serrano (Ed.): *Balance de la Antropología en América Latina y el Caribe* (Mexico City: CRIM-UNAM), 590 p

1994 With Priscilla Stone and David C. Major (Eds.): *Population and environment: rethinking the debate* (Boulder: Westview), 352 p

L. Arizpe, *Lourdes Arizpe*, SpringerBriefs on Pioneers in Science and Practice 10,
DOI: 10.1007/978-3-319-01896-6_2, © The Author(s) 2014

1996 (Ed.): *The Cultural Dimensions of Global Change: an Anthropological Approach* (Paris: UNESCO), 259 p [Published in Spanish: (Ed.): *Dimensiones culturales del cambio global: una perspectiva antropológica* (Mexico City: CRIM- UNAM, 1997), 430 p]

1996 With María Fernanda Paz; Margarita Velazquez: *Culture and Global Change: Social Perceptions of Deforestation in the Lacandona Rain Forest in Mexico* (Ann Arbor: The University of Michigan Press), 115 p [Published in Spanish: *Cultura y Cambio Global: percepciones sociales sobre la desforestación en la selva lacandona* (Mexico City: CRIM-UNAM—Miguel Ángel Porrúa, 1993), 230 p]

2004 (Ed.): *Los retos culturales de México* (Mexico City: Cámara de Senadores—CRIM-UNAM—Miguel Ángel Porrúa), 388 p

2006 (Ed.): *Los retos culturales de México frente a la globalización* (Mexico City: Cámara de Diputados—CRIM-UNAM—Miguel Ángel Porrúa), 627 p

2006 *Culturas en Movimiento: interactividad cultural y procesos globales* (Mexico City: Cámara de Diputados—CRIM-UNAM—Miguel Ángel Porrúa), 368 p

2007 With Cristina Amescua; José Carlos Luque: *Migración y cultura en América Latina y el Caribe: bibliografía seleccionada* (Mexico City: CRIM-UNAM), 217 p (republished in 2010)

2009 *Compartir el Patrimonio Cultural Inmaterial de México: Ritos y Festividades* (Mexico City: Cámara de Diputados—CRIM-UNAM—Miguel Ángel Porrúa), 249 p (2nd edition, 2011)

2010 (Ed.): *Libertad para Elegir: Cultura, Comunicación y Desarrollo Humano Sustentable* (Mexico City: United Nations Development Program), 244 p

2011 (Ed.): *El Patrimonio Cultural Cívico: la memoria política como capital social* (Mexico City: Cámara de Diputados—CRIM-UNAM—Miguel Ángel Porrúa), 178 p

2011 (Ed.): *Compartir el Patrimonio Cultural Inmaterial: Narrativas y representaciones* (Mexico City: Dirección General de Culturas Populares e Indígenas, CONACULTA—CRIM-UNAM), 328 p

2.2 International Reports

1996 *Our Creative Diversity* (Paris: UNESCO). As a Member of the World Commission on Culture and Development and Coordinator of its Secretariat

1998 *World Culture Report,* vol. 1 (Paris: UNESCO). As Coordinator of the Scientific Committee*Informe Mundial de Cultura de la UNESCO, vol. 1* (Paris: UNESCO)

2001 *World Culture Report*, vol. 2 (Paris: UNESCO). As Coordinator of the Scientific Committee*Informe Mundial de Cultura de la UNESCO, vol. 2* (Paris: UNESCO)

2002 *Crossing the Divide: a Dialogue of Civilizations* (New York: Seton Hall). As member of the Group of Eminent Persons for the Dialogue of Civilizations

2.3 Selected Published Articles

1972 "Zacatipan Kinship Terminology: a Dual Approach", in: *Canadian Review of Sociology and Anthropology*, 9, 3: 227–237

1972 "Nahua Domestic Groups: The Development Cycle of Nahua Domestic Groups in Central Mexico", in: *Kung. Magazine of the London School of Economics Anthropology Society*

1977 "Women in the Informal Sector: the case of Mexico City", in: *Signs* (Chicago: the University of Chicago Press), 3, 1 (Autumn): 25–37

1980 "Cultural Change and Ethnicity in Rural Mexico", in: David Preston (Ed.): *Environment, Society and Rural Change in Latin America* (New York, Toronto: John Wiley and Sons): 123–134

1981 With Josefina Aranda: "The Comparative Advantages of Women's Disadvantages: Women Workers in the Strawberry Agroindustry in Agribusiness in Mexico", in: *Signs* (Chicago: The University of Chicago Press), 7, 2 (Winter): 453–473

1981 "The Rural Exodus in Mexico and Mexican Migration to the United States", in: *International Migration Review*, XV, 4 (Winter): 626–650

1981 "Relay Migration and the Survival of the Peasant Household", in: Jorge Balan (Ed.): *Why People Move: comparative perspectives on the dynamics of internal migration* (Paris: UNESCO): 187–210

1982 "Women and Development in Latin America and the Caribbean. Lessons from the Seventies and Hopes for the Future", in: *Development Dialogue* (Uppsala: Dag Hammarskjöld Foundation), 1–2: 74–84

1984 "Agrarian change and the dynamics of women's rural out-migration in Latin America", in: *Women on the Move: Contemporary changes in family and society* (Paris: UNESCO): 83–92

1987 With Carlota Botey: "Mexican Agricultural Development Policy and Its Impact on Rural Women", in: Carmen Diana Deere; Magdalena León (Eds.): *Rural Women and State Policy. Feminist Perspectives on Latin American Agricultural Development* (London: Westview Press): 67–83

1988 "Anthropology in Latin America: Old boundaries, new contexts", in: Christopher Mitchell (Ed.): *Changing Perspectives in Latin American Studies: Insights from six disciplines* (Stanford: Stanford University Press): 143–161

Lourdes Arizpe (Ed.): *Antropología Breve de México* [Brief Antrhropoloby of Mexico] (Mexico City: Academia de la Investigación Científica, 1993)

1988 "Politics of culture as an issue for survival", in: *India International Quarterly*, 15, 3: 69–75

1989 With Fanny Salinas, and Margarita Velazquez: "Effects of the Economic Crisis 1980–1985 on the living conditions of peasant women in Mexico", in: *The Invisible Adjustment: poor women and the economic crisis* (Chile: UNICEF): 242–262

1989 "On the Social and Cultural Sustainability of World Development", in: Louis Emmerij (Ed.): *One World or Several?* (Paris: OECD Development Centre): 207–219

1991 "The Global Cube: Microsocial Models in a Global Context", in: *International Social Science Journal: Global Environmental Change* (Paris: UNESCO), XLIII, 130: 599–609

1991 "A Global Perspective to Build a Sustainable Future", in: *Development*: 6–12

1992 With Margarita Velazquez: "Population and Societies", in: *Global Change and the Human Prospect: Issues in Population, Science, Technology and Equity*

1992 With Wolfgang Lutz; Robert Constanza: "Population and Natural Resource Use", in: J.C.I. Dooge; J.W.M. la Rivière; J. Morton-Lefèvre; T. O'Riordan; F. Praderie (Eds.): *An Agenda for Science for Environment and Development into the 21st Century* (Cambridge: University of Cambridge): 61–68

1992 "Culture and Knowledge in Development", in: Uner Kirdar (Ed.): *Change: Threat or Opportunity for Human Progress? Vol. IV: Changes in the Human Dimensions of Development, Ethics and Values* (Nueva York: United Nations): 117–125

1992 With Fernanda Paz: "Culture et Durabilite, in: *Revue Tiers-Monde: Environnement et developpement*, XXXIII,130 (April-June): 338–355

1993 "An Overview of Women Education in Latin America and the Caribbean", in: J.K. Conway; S. Bourque (Eds.): *The Politics of Women's Education* (Ann Arbor: University of Michigan Press): 171–183

1994 With Margarita Velazquez: "The Social Dimensions of Population", in: Lourdes Arizpe; Priscilla Stone; David C. Major (Eds.): *Population and environment: rethinking the debate* (Boulder: West View Press): 15–40

1996 "Scale and interaction in cultural processes: towards an anthropological perspective of global change", in: Arizpe Lourdes (Ed.): *The cultural dimensions of global change. An anthropological approach* (Paris: UNESCO): 89–107

1996 "Chiapas: The Basic Problems", in: *Identities: Global Studies in Culture and Power*, 3, 1–2: 219–233

1996 "L'innovation et la technologie au service du patrimoine de l'humanité", in: *Actes du Colloque Innovation et technologie au service du patrimoine de l'humanité* (Paris: ADMITECH-UNESCO)

1998 "Sustainable Development in Forest Areas: Balancing Rights and Opportunities", in: Valentina Napolitano; Xochitl Leyva Solano (Eds.): *Encuentros Antropológicos: Power, Identity and Mobility in Mexican Society* (London: Institute of Latin America Studies/London University): 95–109

1998 "Conviviability: The Role of Civil Society in Development", in: Amanda Bernard; Henny Helmich; Percy B. Lehning (Eds.): *Civil Society and International Development* (Paris: OECD Publishing): 21–24

1998 "Our Creative Diversity: A composite view of Culture and Development", in: Susan Wright (Ed.): *Cultural Diversity and Citizenship. Report of A joint UNESCO/University of Bimingham Seminar* (Birmingham: The University of Birmingham): 13–22

2000 With Guiomar Alonso: "Culture, Globalization and International Trade", in: *Human Development Report 1999: Globalization with a human face* (New York: UNDP): 37–56

2000 "Diversity and Integration: Towards a Research Agenda in Latin American Social Sciences", in: Jaime Behar (Ed.): *Inequality, Democracy and Sustainable Development in Latin America* (Stockholm: Institute of Latin American Studies/Stockholm University): 155–164

2000 "Cultural Heritage and Globalization", in: Erica Avrami; Randall Mason; Marta de la Torre (Eds.): *Values and Heritage Conservation: Research Report* (Los Angeles: The Getty Conservation Institute): 32–37

2000 "Transitions in Development: Sustainability, equity and conviviability", in: *Development*, 43, 4: 15–16

2001 "Cultura, Creatividad y Gobernabilidad", in: Daniel Mato (Ed.): *Estudios Latinoamericanos sobre cultura y transformaciones sociales en tiempos de globalización*: 31–48

2001 With Elizabeth Jelin; Mohan Rao; Paul Streeten: "Cultural Diversity, Conflicts and Pluralism", in: *World Culture Report 2000*

2002 "No Alternatives without Diversity", in: *Development* (London, Thousand Oaks, CA and New Delhi: Society for International Development—SAGE Publications), 45, 2: 22–24

2002 "Questions on the notion of museum of civilization (A debate between Lourdes Arizpe and Azedine Beschaouch)" in: *Museum International* (Paris: UNESCO), 54, 1–2: 144–148

2003 With Enrique Nalda: "Patrimonio Cultural", in: Néstor García Canclini (Ed.): *Cultura y Desarrollo* (México: Organización de Estados Iberoamericanos)

2004 "The Intellectual History of Culture and Development Institutions" in: Viajayendra Rao y Michael Walton (Eds): *Culture and Public Action* (Palo Alto: Stanford University Press): 163–184

2004 "Migración y Cultura: las Redes Simbólicas del Futuro" in: Lourdes Arizpe (Coord.): *Los retos culturales de México* (Mexico City: CRIM-UNAM—Miguel Ángel Porrúa—Cámara de Senadores): 19–42

2004 "El espacio cultural global", in: Néstor García Canclini (Coord.): *Reabrir espacios públicos, políticas culturales y ciudadanía* (México: Plaza y Valdés—Universidad Autónoma Metropolituana): 43–59

2004 "Intangible Cultural Heritage, Diversity and Coherence", in: *Museum International* (Paris: UNESCO), 56, 1–2: 130–135

2005 "La transformación de la cultura en México" in: Raúl Béjar; Héctor Rosales (Eds.): *La identidad nacional mexicana como problema político y cultural, nuevas miradas* (México: UNAM, CRIM): 37–55

2005 "Culture, Governance and Globalization", in: *Development*, 48, 1: 35–39

2006 "Mexicanidad, migración y globalización", in: Lourdes Arizpe (Ed.): *Los retos culturales de México frente a la globalización* (México: CRIM-UNAM—Miguel Ángel Porrúa—Cámara de Diputados): 19–48

2006 "Women, power and international ideals" in: Ingeborg Breines; Hans d'Orville (Eds.): *60 women contributing to the 60 years of UNESCO. Constructing the Foundations of Peace* (Paris: UNESCO): 32–38

2007 "The World is Becoming on More Dangerous Place: Culture and identity among Mexican migrants in the United States", in: *Development*, 50, 4: 6–12

2008 "The Ritual and the Promise: Why People Value Social Ritual", in: M. Hutter; David Throsby (Ed.): *Beyond Price: Value in Culture, Economics, and the Arts* (Cambridge: Cambridge University Press): 141–158

2008 "Libertad cultural y redes simbólicas", in: Enrique Florescano; José Woldenberg (Eds.): *Cultura mexicana: revisión y prospectiva* (México: Taurus): 289–324

2009 "El Patrimonio cultural inmaterial y el futuro", in: Maria Angels Ruf Riba (Coord.): *L'anima de la humanitat, el patrimonio cultural inmaterial* (Andorra: Universitat d'Estiu d'Andorra): 103–115

2010 "Desarrollo, patrimonio cultural inmaterial y vida privada", in: Lourdes Arizpe (Coord.): *Libertad para Elegir: Cultura, Comunicación y Desarrollo Humano Sustentable* (México: UNDP): 17–34

2011 "El Simulacro de la Independencia en Chilacachapa, Guerrero" in: Lourdes Arizpe (Coord.): *El Patrimonio Cultural Cívico: la memoria política como capital social* (México: Cámara de Diputados—UNAM—Miguel Ángel Porrúa): 31–43

2011 "El Festejo de la Toma de la Alhóndiga de Granaditas en Tetelpa, Morelos", in: Lourdes Arizpe (Ed.): *El Patrimonio Cultural Cívico: la memoria política como capital social* (México: Cámara de Diputados—CRIM-UNAM—Miguel Ángel Porrúa): 79–87

2011 "Acatempan: el abrazo de la palabra hermana en Guerrero", in: Lourdes Arizpe (Ed.): *El Patrimonio Cultural Cívico: la memoria política como capital social* (México: Cámara de Diputados—Miguel—UNAM Ángel Porrúa): 103–111

2011 "Cultura e Identidad. Mexicanos en la era global", in: *Revista de la Universidad Nacional Autónoma de México* (México: UNAM), 92: 70–81

2011 "Fusión y fricción en la creatividad cultural" in: Lourdes Arizpe (Coord.): *Compartir el Patrimonio Cultural Inmaterial: Narrativas y representaciones* (Mexico City: Dirección General de Culturas Populares e Indigenas—CONACULTA—CRIM-UNAM): 33–44

2012 With Edith Pérez y Luis Miguel Moraita: "Patrimonio Cultural Intangible de Morelos", in: Marcela Tostado (Ed.), *Historia de Morelos. Tierra, gente, tiempos del sur* (Cuernavaca, Morelos: Congreso del Estado de Morelos, LI Legislatura—UAEM—Instituto de Cultura de Morelos—Navarro Editores)

2012 "Interview: Lourdes Arizpe: Mexican Culture: Strong enough to survive in the Era of the Internet", in: *Mexico's States of Opportunity 2012* (México: UNAM—PricewaterhouseCoopers S.C.): 80–83

Part II
On the Author's Selected Major Texts

Discussing the legislative proposals for a Law on Culture in Oaxaca, Mexico in 2008. *Source* From the personal photo collection of the author

Chapter 3
The Dialogue on the Diversity of Cultures and Civilizations

The world is one but the many have not yet found their place in it.[1] Our own nature as human beings makes us forever look at the world from a specific place, a specific time. The horizon of our eyes has always been transformed into the boundary of 'our world'. What happens when we can see beyond our known horizon, to the other side of the world? What we see are countless different cultures that are now part of 'our world'.

The dialogue to relocate 'ourselves' and countless 'others' in this Bold New World (Knoke 1996) has only just begun. Once the terms are clear a new cartography of cultures must be developed. Yet something vaster is happening that any dialogue must endeavour to shape. A new consciousness is on the rise: people in every culture are becoming keenly aware of their singularity and yet painfully aware of the need to empathize with their fellow cohabitants on the other side of the world.

Could we not aspire, in this new millennium, to extending this new world horizon through empathy with no boundaries, imagination with no barriers, and creativity with no restrictions? We can aspire to it, certainly, but the more the basic needs of so many poor people are not met, the more resentment grows, the more conflicts will erupt into wars. However, let us not lose sight of our purpose in this horizon.

As the report *Crossing the Divide*[2] states, it is the heightened degree of interdependence in the world that has transformed any 'threat' into a 'global threat' that

---

[1] Paper delivered at a seminar of the International Union of Anthropological and Ethnological Sciences and the International Social Science Council on 'Research on the Diversity of Cultures', Paris, June 2001. This text is unpublished.

[2] Lourdes Arizpe was a member of the 'Group of Eminent Persons for the Dialogue among Civilizations' created by the United Nations in 2000, year of the 'Dialogue among Civilizations'. Together with Giandomenico Picco, A. Kamal Aboulmagd, Hanan Ashrawi, Ruth Cardoso, Jacques Delors, Leslie Gelb, Nadine Gordimer, Prince El Hassan bin Talal, Sergey Kapitza, Hayao Kawai, Tommy Koh, Hans Küng, Graça Machel, Amartya Sen, Song Jian, Dick Spring, Tu Weiming, Richard von Weizsäcker, and Javad Zarif, Lourdes Arizpe wrote the report *Crossing the Divide: Dialogue among Civilizations* (2001) which was presented to the UN General Assembly on 8 November 2001.

L. Arizpe, *Lourdes Arizpe*, SpringerBriefs on Pioneers in Science and Practice 10,
DOI: 10.1007/978-3-319-01896-6_3,

knows no boundaries. What the recent tragedies in New York and in Afghanistan have demonstrated is, on the one hand, that certain kinds of violence have acquired a new global rank but, on the other, that so has the collective will for peace. It is urgent then to ask what kind of an international social contract must be built, urgently, to allow humanity not to fall into irreversible unsustainability.

Just a few days ago, a student asked me how we were going to build a transition to democracy, a dialogue of cultures, in a 'world at war'. This is precisely the kind of false perception that must be shattered. The world is not at war. It is groups with particular interests that are holding the world hostage. Yet the background of the brutal assassination of so many people, in the World Trade Center in New York and now in Afghanistan, requires a rethinking of the work to be done: by international organizations, by governments, by scholars, by conscious and active citizens of the world.

The terrorist attack of 11 September 2001 has highlighted how, through the cultural and religious dimensions of such conflicts, groups of extremists can have a far greater impact than their political or intellectual resources would otherwise make possible. Yet a reaction of violence against violence creates a spiral of suffering that will lead, unforgivingly and inevitably, to more violence.

Discussing intangible cultural heritage at a municipal meeting in Zacualpan, Morelos, Mexico in 2007. *Source* From the personal photo collection of the author

An initial reaction to these tragic events gives the impression that the international work carried out in previous years on cultures of peace, the relationship between culture and development, and on the dialogue of cultures has been insufficient. Its contribution, however, was reflected in the many statements and demonstrations showing that the vast majority of people in the world are on the side of peace and of cultural coexistence. *Perhaps never in history has this collective will against violence been so evident and so global in its manifestations.* This, I believe, is the movement we must act upon through concerted international programmes and actions.

These tragic events, if anything, demonstrate how necessary policies and programmes related to culture have become on the world stage. No longer will matters of cultural identity or religious fundamentalism be considered a marginal concern in international geopolitics or in the defence of democracy.

Since the 1960s UNESCO has had the foresight to call the world's attention to the importance of culture and cultural policies in development. At present, much more international cooperation and analytical work is needed to understand how a person's economic, political and social conditions interact with elements of cultural behaviour.

3.1 Cultures or Civilizations?

At this present time, even the vocabulary for conducting this dialogue of the many is still being constructed. Are the patterns of significant cultural behaviour that people express and represent today 'cultures' or 'civilizations'? In a way, this is a misleading debate. Firstly, as the President of the Council of Europe expressed it at the General Assembly of the United Nations, "the dialogue of civilizations cannot be reduced to a dialogue of religions". Civilizations, in the classical historical sense, rose as a centralized state organized economic and political forces, fostering the specialization of functionaries, artisans, merchants, warriors and artists. As a result, commerce, engineering, architecture and the arts flourished, bestowing upon that particular civilization specific historical achievements.

Secondly, however, such functional specialization also led to cultural diversification so that every great civilization developed, ipso facto, in its midst, other civilizations. This is certainly true of those civilizations talked much of today, for example, the West and Islam, but also of the Indian and Chinese civilizations. In fact, as many critics of Huntington's *Clash of Civilizations* (1996) have argued, most wars in the last few centuries have occurred *within* the boundaries of these major cultural units.

Thirdly, all civilizations, *at different scales*, are transversally linked through languages and 'borderline' cultures or, as the more contemporary term expresses it, hybrid cultures—for example, Christian Copts, or Bengali Muslims, or Catholic Amerindians, or Japanese Brazilians. In other words, such cultures can only be called 'borderline' if such a border is artificially imposed on boundaries which are permeable, are indefinable by anthropological standards, and shift almost in every generation. A recent example has been the debate about whether Islamic cultures are part of the West. Indeed, one can go so far as to ask whether there is any sense in trying to work out whether cultures that have been in contact for ten thousand years in Eurasia, and have enriched each other constantly throughout that time, can be defined either as 'civilizations' or as 'cultures'.

Evidently they can be in political debate, for political purposes, and this is the fourth point. Anthropology discarded the term 'civilization' as an analytic term back in the 1950s because it introduces hierarchy, it flattens internal cultural diversity, and it insidiously introduces notions of unilinear evolution within the borders of the groups claiming to be civilizations over and above their cultural constituencies. Indeed, would it not be a pyrrhic victory if the demand for a dialogue of civilizations, set forth with a desire to defend diversity on the world stage, were to become a way of denying the richness of the diversity of civilizations and cultures within the macro-scale civilizations? Would it not be a way, again, of centralizing the representation of power so that only a few players are seen to be deciding on the fate of the world? If so, this would go against the world trend towards democracy and the participation of the many.

Given these concerns, how can the Dialogue of Cultures and Civilizations make a significant contribution to world communication and understanding among peoples?

3.2 A Higher Order of Civilization

Many writers in the aftermath of the attacks have asked whether the recent conflict is really a 'clash of civilizations', as expressed in the popular rendering of Huntington's thesis in the media. In a play on words, some authors have posed the question of whether in fact the current conflict reflects a '*crash* of civilizations'. Especially, as one writer put it, each side labels the other as 'The Devil'.

Indeed, what kind of civilization tolerates fanatics that kill children, women and men? And, again, what kind of civilization mobilizes its armies against a whole country to punish rulers it has supported for so long and ends up killing children, women and men?

Which civilizations are these that no human being, whether a true Muslim, Jew or Christian, would recognize as their own? They would not be able to identify them because they all recognize *a higher order of civilization*, one that has been created and nurtured by the combination of many, many strands of philosophical and theological thinking throughout human history. As eloquently expressed by Abldelaziz Belkhadem, foreign minister of Algeria, at the session on the Dialogue of Civilizations at the UN "No one can doubt that we are witnessing the more and more extensive formation of a civilization of the universal which is the result and the fruit of the endowments and contributions of different human civilizations since the night of time".

The outcry around the world against the recent and continuing killings of innocent people, whether in New York, Tel Aviv or Kabul, is a vivid exemplar that makes explicit the main principle of this world civilization. That principle is, that every individual respect the life of others, in the same way that he or she desires to be treated, with dignity, empathy and tolerance. In *Crossing the Divide* (Picco/Aboulmagd/Arizpe/Ashrawi/Cardoso/Delors/Gelb/Gordimer/bin Talal/Kapitza/Kawai/Koh/Küng/Machel/Sen/Jian/Song/Spring/Weiming/von Weizsäcker/Zarif 2001) this tenet is noted as the Kantian principle that is reflected in very similar precepts found in so many cultures and religions. Hans Küng eloquently expressed it at the UN session: "In this dialogue the world's religions have rediscovered that their own fundamental ethical teachings support and deepen those secular ethical values which are enshrined in the Universal Declaration of Human Rights."

It is this higher order of civilization that must be built through the 'global ethics' proposed by the United Nations World Commission on Culture and Development in *Our Creative Diversity* (UNESCO 1995). The Commission posited that human rights, democracy, equity and sustainability are the principles on which a civilized global society must be built. For this civilized global society I propose the term *geopolis*, from the Greek *geo* meaning Earth and *polis*, a community whose members willingly agree to abide by certain rules that bind them in solidarity to each other.

It is becoming more and more urgent to build this *geopolis* now that contemporary societies face a complexity which is different from anything we could learn from in history. New values and new standards must be forged and, for this, as I noted at the conclusion of UNESCO's 1998 Intergovernmental Conference on Cultural Policies for Development, we need 'freedom to create'. Creativity and not only a return to traditions is what is needed to shape this new *geopolis*.

3.3 Cultural Diversity in the *Geopolis*

Freedom to create goes hand in hand with respect for and appreciation of cultural diversity. As I expressed it at a New York University meeting on Culture and Trade in March 2000—taken up in the UNESCO *Declaration on Cultural Diversity*: "Diversity is the source of human capability of developing: we think by associating different images; we identify by contrasting ways of living; we elect by choosing from an array of options; we grow by rebuilding our confidence again and again through dialogue".

The UNESCO *Declaration on Cultural Diversity* will have a momentous impact, precisely at a time when the world needs a consensus among world governments and civil society commissions and organizations on the way forward in building a socially sustainable future. In this new beginning, to cope with the momentous challenges of sustainability, governance and *convivencia*[3] in a global era, we need cooperation on a world scale, putting into play all the creativity that be summoned from all cultures and civilizations.

As was explained in the Second World Culture Report "it is no longer a matter of globalization allowing cultural diversity to continue to develop, it is cultural diversity as a condition without which globalization cannot continue".

[3] *Convivencia* in Spanish means not only the conviviality of sharing an agreeable feast as in English or French. *Vivencia* in Spanish means a life experience so that *convivencia* means not only sharing together but actually living the experience together.

Diversity must also include all the diverse sectors of societies. Most particularly, women. Civilizations have been built by men and women, each with their respective and complementary contributions. No dialogue among civilizations could take place without the active and inventive participation of women.

Respect and reciprocity cannot be decided by law or imposed by institutions, although disrespect and hierarchy can be and often are. Minimizing inequality in the social primary goods in Rawls's sense—not just rights and liberties but also powers and opportunities, income and wealth, and the basis of self-respect—is not only the most effective instrument in this regard but can also be institutionalized. Minimizing inequality, not just absolute poverty, empowers people with the possibility of equal and effective participation and, thereby, of *convivencia* and genuine cultural pluralism.

The importance of the UNESCO *Declaration on Cultural Diversity* is that it gives content to the Dialogue of Cultures and Civilizations. It also dissipates the false problems in the definitions of cultures and civilizations by placing *diversity* and *creativity* at the centre of world cultural development.

3.4 Relocating Cultural Diversity

In a shrinking planet, wired and webbed by the most continuous and interactive cultural contacts in history, people are constantly having to negotiate with others who have different values, attitudes and behaviour. As explained in the Second World Culture Report, such necessary transactions become impossible if a cultural canon is elevated to the level of a metaphysical condition. Questions of faith, all would agree, cannot be negotiated. If so, groups defining themselves *exclusively* on the basis of religion, at best, can only negotiate coexistence and tolerance, and at worse, as Osama Bin Laden has declared, they will fight to the death in a 'war of religions'.

Democracy, trade and policy-making, on the other hand, imply negotiated conciliation, as do civilized international relations. The only way to achieve this conciliation, as Umberto Eco has reminded us, is to consider that "… all wars of religion that have bloodied the world for centuries have been born out of passionate adherence to simplistic oppositions: We and They, the Good and the Bad, White and Black. If Western culture has shown itself to be creative…it is because it has striven to 'dissolve' nefarious simplification with the light of critical spirit and enquiry."

Simplifications of political philosophies do not generally last very long: freedom of expression and open debate lead to adjustment through accurate criticism and to negotiation between contending parties.

In his statement at the UN General Assembly, the European Union spokesman declared that "the right to difference and to identity is inseparable from that of the equal dignity of cultures. It is within this perspective that the European Union conceives the Dialogue among Civilizations. This must be carried out without taboos. We have the right to ask questions to a civilization other than our own and to pose questions to others. Others also have the right to interrogate us on the reason for certain inequalities. If the Dialogue of Civilization is to be reduced to meeting to congratulate ourselves, it is not a dialogue."

In *Crossing the Divide* (Picco/Aboulmagd/Arizpe/Ashrawi/Cardoso/Delors/Gelb/Gordimer/bin Talal/Kapitza/Kawai/Koh/Küng/Machel/Sen/Jian/Song/Spring/Weiming/von Weizsäcker/Zarif 2001) we agreed, precisely, that the framework of international relations must now change so that the Other is seen no longer as an Enemy but as a competitor in creating the best for improving the lives of all. To do this, the *geopolis* must be filled with narratives that may be criticized, dissected, remodelled and transformed until new consensual narratives are put into place.

3.5 Equality of Vulnerability

The urgency of conducting a Dialogue on the Diversity of Cultures and Civilizations must also be linked to a sustainable future, both socially and ecologically. Scientists who met at the World Science Organization's "Open Conference on the Challenges of a Changing Earth" in Amsterdam in July 2001 confirmed that global warming will have decisive impacts on the life of every inhabitant of the planet. Global environmental change thus creates an *equality of vulnerability* which has also deepened through increased interdependence within a single world economic system.

In *Crossing the Divide* we proposed that equality in vulnerability heightens the need for a broader, more political dialogue among cultures and civilizations. Thus, it stimulates dialogue. For the real answer to equality in vulnerability, which could lead to equality of opportunity, is adherence to accepted forms of common behaviour by more and more actors on the international scene. In the report, we agreed that this would require that every single decision of individual members of the international community must be taken into account. (Picco/Aboulmagd/Arizpe/Ashrawi/Cardoso/Delors/Gelb/Gordimer/bin Talal/Kapitza/Kawai/Koh/Küng/Machel/Sen/Jian/Song/Spring/Weiming/von Weizsäcker/Zarif 2001).

The report goes on to say that "perhaps what we are really talking about are no longer individual enemies for individual countries but a multifaceted enemy for all. The spreading of contagious disease, weapons of mass destruction, unrestricted dissemination of small weapons, poverty, all represent different faces of an 'enemy' for the entire human race.... If the enemy is common, it follows that fighting against it requires unanimity" (Picco/Aboulmagd/Arizpe/Ashrawi/Cardoso/Delors/Gelb/Gordimer/bin Talal/Kapitza/Kawai/Koh/Küng/Machel/Sen/Jian/Song/Spring/Weiming/von Weizsäcker/Zarif 2001). How do we begin to build such unanimity of purpose?

3.6 The Need for a Social Science Dialogue

Unanimity of purpose can only come from sustained, informed, conceptually rigorous debate that only analytic social science can provide. And we must ask why political and social creativity has stalled so noticeably in the last two decades.

At the end of the nineteenth century, the development of industrial capitalism in European countries had exacerbated social problems such as the disintegration of families and the abandonment of children; labour problems of low wages and unhealthy work conditions; problems of criminality, alcoholism and violence against women; and cultural dislocations. As explained in the *Social Science Report*, to deal with this 'social question', governments at that time created and supported research in the social sciences which contributed to policies for improving the lives of people in the new industrial and urban settings. Today, when precisely similar problems can be perceived in vastly greater numbers in so many countries, the prevailing policies in most developing countries emphasize administration and engineering to the detriment of funding for social science research and reflection.

One of the results of such measures is that people find no contending ideas they can debate, reject or adopt that will give them a feeling that they have a say in constructing new meanings to their changed lives. They thus feel disenchanted in their desire to forge their own futures, or they face a void of meaning that makes them turn to nefarious ideologies.

Ruth Cardoso, a distinguished Latin American anthropologist who was also a member of the Group of Eminent Persons for the Dialogue among Civilizations, emphasized in her speech at the UN General Assembly the fact that cultural studies and social movements of mobilized peoples of African descent, women and peoples of varied ethnic and social origins, in many countries, contributed greatly to the political evolution of the end of the twentieth century. Unless all such movements are incorporated into the Dialogue of Cultures and Civilizations, the world will face the challenge that Cardoso has also called attention to, that is, "the construction of identities so strong as to exclude the principle of multiculturalism and the fight against discrimination, and the restructuring of values and patterns of behaviour leading to intolerant fundamentalisms".

3.7 A New Collective Will

The Dialogue of Cultures and Civilizations must lead to the careful and conscious building of a new constellation of political will. UNESCO, which has both the experience and the commitment, as was confirmed in the last General Conference, must lead the way.

For this Dialogue to be successful, a different concept of the role of the United Nations is needed. As proposed in *Crossing the Divide*: "To those who say that the

United Nations is nothing more than the sum of its members, we beg to differ. Within the framework of the United Nations, we would like to submit that an international social contract is being consummated....For the need for such a contract will become more and more self-evident, as power alone will not deliver peace any longer" (Picco/Aboulmagd/Arizpe/Ashrawi/Cardoso/Delors/Gelb/Gordimer/bin Talal/Kapitza/Kawai/Koh/Küng/Machel/Sen/Jian/Song/Spring/Weiming/von Weizsäcker/Zarif 2001).

The sentiment I would like to end this essay with is that there is no 'global threat' that our deliberate, collective will cannot overcome.

References

Huntington, Samuel P., 1996: *The Clash of Civilizations and the Remaking of World Order* (New York: Simon & Schuster).

Knoke, William, 1996: *Bold New World* (New York: Kodansha).

Picco, Giandomenico; Aboulmagd, A. Kamal; Arizpe, Lourdes; Ashrawi, Hanan; Cardoso, Ruth; Delors, Jacques; Gelb, Leslie; Gordimer, Nadine; bin Talal, Prince El Hassan; Kapitza, Sergey; Kawai, Hayao; Koh, Tommy; Küng, Hans; Machel, Graça; Sen, Amartya; Song, Jian; Spring, Dick; Weiming, Tu; von Weizsäcker, Richard; Zarif, Javad, 2001: *Crossing the Divide: Dialogue among Civilizations* (New Jersey: Seton Hall University).

United Nations Educational, Scientific and Cultural Organization, 1995: *Our Creative Diversity* (Paris: Culture and Development Co-ordination Office).

Other Literature

United Nations Educational, Scientific and Cultural Organization (adopted 2 November 2001), *Declaration on Cultural Diversity*, UNESCO Records of the General Conference, 31st Session, Volume 1: Resolutions, Paris 15 October to 3 November, UNESCO Legal Instruments; at: <http://unesdoc.unesco.org/images/0012/001246/124687e.pdf#page=67> (12 May 2013): 1.

Chapter 4
On the Cultural and Social Sustainability of World Development

To examine the issue of interdependence in a multipolar and globalised, yet dualistic world economy, we must start by clarifying the frame of reference in which it is to be analysed. As a starting point, all we have learned through historical and anthropological research indicates that the world today is facing not only a rearrangement of economic polarities, but a full-scale civilizational change.[1]

Such change, as in previous historical instances, is being driven by technological advances that will alter the mechanics of everyday life, as well as the social structure based on forms of production organised on the raw materials-factory-assembly line pattern of nineteenth century industry. Policy changes stressing the market economy and economic globalization of the economy are also bringing about a political reordering of capitalist and socialist industrialism. The changing role of the State and the challenge of ethnic and cultural minorities, in the context of trends towards global interdependence, will require a redefinition of the political structure of nation states. Biotechnology, in a few years, will change the nature of agriculture, which is still the basis of the livelihood of the majority of the world population especially in countries of the South. At the same time, human biogenetics and demographic trends have already impinged on the most private of domains, that of the family and of biological and social reproduction. Post-industrial culture, which is surely a misnomer, in the North is fostering a fragmented sense of self and of social place. In developing countries people perceive their traditional cultures and identities as alarmingly imperilled by the new possibilities of electronic communication and audiovisual culture. Other changes can be predicted as well in other social domains. The main point, though is that there have been only two previous instances in history in which such overall social change has taken place, that is, in which all areas of human and social life are being affected. It is, however, the first time in history that the whole of the world is

---

[1] This text was first published as: "On the Cultural and Social Sustainability of World Development", in: Emmerj, Louis, (Ed.): *One World or Several?* (Paris: OECD Development Centre, 1989): pages. Republished in: *Development*, 1997: 43–57. The permission to republish this text was granted by OECD on 22 May 2013 (Reference P-2013-179).

L. Arizpe, *Lourdes Arizpe*, SpringerBriefs on Pioneers in Science and Practice 10,
DOI: 10.1007/978-3-319-01896-6_4, © The Author(s) 2014

being affected almost simultaneously. However, as Soedjatmoko perceptively states, the Third Technological Revolution is already underway before the First and Second Technological Revolutions have gotten to the whole globe.

In spite of these two clearly defined characteristics of today's global change, though, discussion of world trends has not been holistic in analysing and in proposing policy actions. It has concentrated on finding solutions to economic problems, and on beginning to create awareness as to the ecological sustainability of development actions. This paper will stress that both the economic and the ecological challenges are interwoven to issues related to the social and cultural sustainability of world development.

Secondly, discussion of world trends has not been global in the sense of trying to create a world citizenship, instead of trying to accommodate as far as possible national, corporate and conglomerate economic interests. As an unnamed Danish poet—quoted by Inga Thorsson—said, we are still global citizens with a tribal mind.

If we are facing a major civilizational change, should we then concentrate on creating a 'grand design' for the future? This is neither possible nor desirable. History shows that all great social transformations have happened without a blueprint- or most of them, at least. And changes today are so varied and so inextricably linked that unidisciplinary predictions are hardly worth the effort. However, it is certainly true that human goals and shared perceptions have guided such transformations and, especially, have given people a sense of purpose and of solidarity without which changes can become dangerously violent. Thus, we must at least begin to think of policy guidelines that can be propitiatory actions for a new world society.

4.1 A First Step: Global Accountability

While we cannot predict where change is leading us, steps must be taken to ensure greater accountability of national governments and multinational conglomerates to the global citizenship, in order to stop corporate and national virtues from turning into global vices. As Mahbub Ul Haq has correctly pointed out, just when we require international institutions and internationalism, the United States, World Bank, IMF and GATT have declined in world economic management. One way of doing this would be to develop international law into a global juridical system and to have multilateral institutions slowly evolve towards institutions of a world government. This, of course, will be fiercely opposed by the hegemonic powers but, faced with the globality of issues and the threat of collective survival; it will have to be done sooner or later. Surely some leaders, many thinkers and many global citizens are already willing to break the gridlock of parochial interests to build a new world.

Perhaps this process would be made easier in a one-world approach, that entails more direct negotiations which would allow checks and balances to be introduced and means found to make them effective. In a multipolar scheme of greater autonomy of regions, global negotiations may be more balanced but would regions internally have a pluralistic scheme of negotiation? As many speakers pointed out in the seminar, the North-South divide can also be found within countries, both in the North and in the South.

What should be obvious from previous historical experience is that economic polarization into a two-track system tends to create political, social and cultural tensions that make it unsustainable in the long term.

The Latin American experience with dualistic economies showed them to be inherently unstable. Economic growth in the region since the 1950s created such disparities in income distribution that, among other things, guerrilla warfare became widespread in the sixties. Dictatorial or authoritarian regimes further exacerbated political militancy and had to give way either to military regimes and 'dirty wars', or to costly populist policies; the 'miracles' for the few turned into nightmares for all.

Dualistic economies will be even more unstable internally in the 1990s. First, a higher educational level of most of the population in most developing countries will make economic polarisation politically more untenable. Second, the mass media, unless it is totally reality-washed, will make economic disparities constantly known and thus spark more frustration. Third, at least two of the major religions in the South, Catholicism and Islam, are doctrinally opposed to extremes of wealth and poverty; a dualistic economy will strengthen fundamentalisms, or irrational pseudo-religions, or, in the best of cases, politically conscious theologies such as the Theology of Liberation in Latin America; religious conflict will then fuel political and social strife.

4.2 A Second Step: Taking Social Sustainability into Account

It is well-known, that, since the fifties, development policies have either ignored or taken the social and cultural dimensions of development for granted. There was an implicit assumption that these dimensions would, however muddled, fall into step with whatever new economic realities arose. This was in fact the case until the sixties because in the North, as well as the South, the private sphere of family, reproduction and personal relationships was as stable as the rock of Gibraltar; also, in the South, geographical mobility was running a slow, manageable course; in both hemispheres cultural identities were secure and ethnicity dormant; and in the hegemonic cultural and artistic world in the North, the avant-garde was busy voluntarily delinking itself into an 'art for art's sake' solipsistic spiral. Thus, the train of economic growth was able to run smoothly along the two taken for granted rails of social and cultural stability; by the early seventies, though, in both hemispheres, for different reasons, the train started derailing.

The inertia of the private sphere exploded: some attribute it to the State wanting to control the last autonomous frontier in society; others to the fact that women finally rebelled. It has been argued that the demise of religious belief undermined psychological and moral certainties and created ‘homeless minds’ or ‘hedonistic values’ in the North and messianic—political or religious—movements in the South; that the weakening of traditional religious, community and family institutions left individuals alienated and alone; that the need to control population growth and the invention of the ‘pill’ and of other contraceptives catapulted the whole process of biological and social reproduction into the public sphere; that feminism turned this on-going process into a visible ideological and political movement: in fact, the feminist slogan of ‘the personal is political’ is to date the most succinct description of this historical development. At present, research has shown, though, that this overall change was due to the historical combination of profound and simultaneous changes in religion, reproduction and women’s roles.

Previously, industrialism, both in its capitalist and socialist forms, had found a balance in keeping the private sphere of reproduction on a different logic from that of economic production; now this sphere—both in it’s biological and social dimensions—is posing fundamental and unexpected problems for economic growth. Just to mention the main ones: the decline of population in countries of the North; overpopulation—as related to economic capacity to sustain a given population—in many regions of the South, linked to rapid ecological depletion; continuous South-North labour migrations; the changing gender dimensions of the structure of employment; women’s expanding leadership of urban and rural movements of the poor; growing matrifocality of families, which has also been termed the ‘feminisation of poverty’.

To our dismay, such issues are most often absent from economic development plans, or are mentioned alongside as ‘social problems’. This represents both a theoretical and a methodological challenge for economic development models. This challenge becomes even more important if the problems of blatantly ‘anti-social’ phenomena are added to the picture. If you put together a few ‘hard’ data such as, the fact that US prisons are now insufficient to hold the swelling criminal population, that suicides continue to be one of the major causes of deaths in the North, that drug addiction and production is alarmingly on the increase, that religious or sectarian fanaticisms are growing everywhere, that alcoholism, prostitution and criminality (including militarily sanctioned criminality) are also on the rise globally and that a surprising number of urban people are really believing that, as the Apocalypse says, we are on the verge of millenary destruction, a mentally and socially unbalanced world comes blatantly into view. In strictly economic terms, one could say that the world economy is becoming blacker by the day. Has this been taken into account in development models? Or is this the part that everyone will try to leave out and throw into the hands of the police and the military? This could bring us to a very sombre thought: if part of political sovereignty is being handed to the market, and an increasing part is being handed to the police and the military, how much civil political sovereignty will there be left for development to continue to be a rationally driven activity?

The above has been described in some detail because the whole point of this paper is to stress that, together with the ecological sustainability, priority attention must be given to the social sustainability of economic growth. It should be fairly clear by now that in some vital areas social and cultural factors, such as the ones mentioned above, are already altering the patterns of economic growth.

In this dimension of 'social problems' one must add, in the South, the most massive and disruptive transfer of people from rural areas to urban centres in human history. The social and cultural dislocation which has ensued is unparalleled, although I for one am convinced that human creativity and sociability could rebuild cultural ties in a few generations. That is, if conditions such that allow it and this should be one of the most crucial areas for policy action in the coming decade.

For in speaking about the population which is in danger of being involuntarily delinked or marginalized in a two-speed world economy, let us be very clear who we are speaking about. Without doubt we are referring mainly to rural peoples, farming families mostly, and the urban poor which, in the South, are overwhelmingly migrant farmers. Indeed, farming families will probably be the most victimized social group in the globe in the coming decades, whichever model of world economy is taken, since they are caught between three advancing bulldozers. The first, biotechnology, is about to change a form of production which is 15,000 years old, into an unrecognizable new thing. In this new thing, farmers will not be able to farm without the high technology inputs controlled by the few. Never has a power of such extraordinary implications been held by the few. The second is linked to crash of agricultural commodities in the world market which is making farmers all over the globe go bankrupt. Those in the North, however, have rich governments who subsidize, regulate, give Social Security, turn farms into hotels or otherwise find ways of minimizing the blows to farmers. In the South, farmers are blown away like the sand in the desert, thus creating a social desertification of rural areas just as damaging as the soil desertification. And so, the third bulldozer is hitting harder in the South as well, and this is ecological depletion. The combination of economic, biotechnological and political factors are taking their livelihoods away from millions of farming families, aggravated, in some cases, by demographic trends, who are now edged towards depleting ecological resources. Surely this must be a high priority area for policy action in the coming decades. But not by way of providing only immediate relief; the challenge is to rethink the economy of the future in terms of a unified territorial distribution in which the rural form of production can be linked to the high speed on. Further research and creative thinking, as well as political will, are needed to achieve this.

4.3 A Third Step: Cultural Sustainability

The kind of civilizational transformation the world is undergoing seems to entail a homogenization process, one leading towards a 'global village'. It is true that all over the world there is alarm about the possible 'loss of identity', whether in

France, in Mexico or in Singapore, as the mass media sweeps over continents erasing our 'sense of place'.

However, in spite of a superficial trend towards cultural homogenization, or, more likely, because of it, never has cultural diversity been so often, and so belligerently, defended in the modern era. As centripetal forces at the global level pull towards a transnationalization of the economy and of culture, the reverse seems to be happening at the micro-social level as centrifugal forces atomize national identities into ever smaller units. The main issue here for development is how a two-speed world economy or a one-world model would interact with the boundaries of national, ethnic or communal identities.

On this issue it is important to be aware of the following analytically different processes:

In many countries, some groups, consciously and willingly, will want to stay out of a modernizing development, for the time cycles in which they live are circular, and, in fact, timeless. And their wishes must be respected. This will be the case of some religious sects, and some ethnic groups, both in the North and in the South. The decision should be left to them, with a clear understanding that they will then not desire the goods and services of a modernizing economy. Their problem may be, then, that many of their younger generations will constantly be slipping out to the consumer society. But this will be their problem, just as their standards of living will be their choice.

If, instead of that they are forcibly or charitatively handed out a Western way a life, their psychological and symbolic structures will be destroyed as has happened with indigenous peoples all over the world. What they do need is State protection against those who covet their lands, or who want to exploit their labour. New schemes to do this are now being explored in different countries: in Canada, the Inuit are negotiating that a separate Province be created for them in the Arctic lands; in Nicaragua a statute of regional autonomy was granted to the Atlantic coastal area where most Indian and black minorities live.

The main point is that these groups are fighting because they don't want to change. But what will happen is the basis of their livelihood is changed, even indirectly, by biotechnology, by environmental concerns or by telecommunications? They may then fall into a second category.

That first group, who are defending an ancient form life and culture, must be distinguished from those who are retrenching back into traditional customs or religions because they are being marginalized. That is, they lose their livelihood, or they foresee that the new economic context will be barren land for them, and this may be quite factual since in a dualistic economy they would most likely be left in the stationary side. This would be the large majority of the population in rural areas in the South, as well as unskilled workers in the cities, both in the South and in the North. It seems to me that it is this perception of their increasing exclusion from the high-speed economy, usually accompanied by political marginalisation as well, that leads them to take up sectarian identities, be they religious, esoteric, political or ethnic. Because they are closer to the high-speed economy in Northern cities, such phenomena are more prevalent there. In the

South, centuries old identities still provide this shelter. The main point here is that the first group is fighting because they don't want to change; the second is reacting because they have been left out of change. If the first group is offered a place in economic growth, they will refuse it; the second will accept it because they have nothing else to fall back on. From a policy perspective it is this very diverse group which requires actions to stimulate their own development but with some way of linking it to the high speed economy to avoid its involuntary delinking from the latter.

The third factor creating centrifugal tendencies which is derived from the process explained above but which is useful to distinguish analytically, is political advantage. Cultural identities have been and are being used to mobilize for liberation purposes, or to gain power in national politics, or to rally macro-regional political forces. Nationalism—both in nation states, or in 'nationalities' in socialist systems—and ethnicism are now being taken up as banners in this struggle. In some cases, groups are demanding greater political autonomy in a most varied range of movements—which it would be too long to go into here—i.e. the Irish, the Basque, the Armenians, the Sikhs or the Caledonians. In other cases, they are combining religious defence with regional influence as in the case of Iran and Libya. The main point here is that, in contrast to the first group, cultural identity is not an end in itself, but a means towards achieving some political purpose.

Finally, centrifugal tendencies are also put into play when cultural identity is used to defend national markets. When the British government launched its 'Buy British' campaign at the end of the sixties, it opened the way for a wide range of protectionist campaigns in commercial warfare among rival economies.

These four different trends must be kept in mind, then, when asking whether plurality and diversity will be part of the new world organisation. It certainly will be. Against those who argue that more development and more mass media automatically mean more cultural homogenization, the best contradicting example is that of Europe itself: in spite of its advanced industrial development and wide-ranging communications—3,000 languages are still spoken there—and the voices of cultural minorities are very much alive.

In the near future, the above four processes will most likely continue to influence social and political events, both as an action as well as a reaction to macroeconomic centripetal tendencies. A two-speed economy, though, would certainly exacerbate such conflicts because marginalisation will most probably run along pre-existing national and ethnic boundaries, thereby deepening disparities of development. This, then, would be a formula for chronic political instability and social strife in all countries.

From the policy perspective, the priority in this cultural dimension would be to change legislation or to create a new one, especially in countries of the South, to ensure a balanced political participation of all ethnic groups, and to give autonomy, as far as possible, to regions and groups who demand it. However, some minority leaders are even demanding that the organization of nation states be dismantled; and this is echoed by those who see the new geo-economic blocs as the major political actors on the world scene. In my view this is both unrealistic and undesirable. All scientific studies have shown that social and cultural diversity is an inherent trait in the human species -in animal species as well- and that only a few generations of relative isolation are necessary for a new culture and social organization to be formed. But the demographic growth patterns, inventiveness and aggressiveness, to name a few traits, of human groups, vary greatly and thus, rules of political organization of a medium range geographical span are necessary to attenuate and solve conflicts. This is what nation states in fact try to do. But this means that the State must be responsive to the demands of cultural minorities. Also, macroeconomies in a multi-polar world would not be able to deal effectively with a myriad minority groups, nor would the latter have any leverage for defense or negotiation in global developments.

4.4 A Special Issue: Knowledge to Sustain Development

Another area which is vital for development, especially in the South, but which routinely gets lost in the interstices between technology and ethnicity, is that of knowledge. I'm referring specifically to that knowledge variously called traditional local, empirical or ethnoscientific, but which is vital to development.

In this respect, it is important to emphasize that the present conditions of the international system are making people in countries of the South poorer in economic resources, but they are also making them poorer in knowledge, and, what is perhaps even worse, poorer in the confidence with which they could continue to create knowledge. This trend will, indeed, foster a historically unprecedented skewness in cognitive and intellectual wealth in a world which has had a multipolar intellectual and scientific history: Egypt, Mesopotamia, India, the Inca, Mesoamerica, China, Benin, the Arabs, Greece, and Europe. Never in history has any region been the only perpetual creator of knowledge.

For although economic multipolarity is envisaged for the future, isn't scientific and technological monopolarity insidiously building up? But let us not confuse the issue: it is not a question of not recognizing the extraordinary achievements of Western science and culture. There is no doubt that the West, through rationalism, created the intellectual conditions to sustain scientific progress. But it is a question of putting it into focus, its great Age of Discovery—which everyone is anxious to celebrate in its 500 anniversary in 1992—, was also its Age of Appropriation of the world's resources. Everyone immediately thinks of economic resources, rarely do they consider the enormous intellectual resources which Europe also took from the East, the South and everywhere.

For, ever since the Elizabethan Age—and some even believe this was one of the crucial inventions in the advent of science and capitalism—European scientists and intellectuals have been scrupulously protected in their creative rights and rewarded in their time-consuming efforts. Compare that to the fact that the 'folk' discoverers and intellectuals of the South have never had the least recognition of their rights or efforts, and this continues right up to the present day.

Corporations and scientists in the North demand that patents should be respected in countries of the South. But where are all the patents for all the man/woman-made products, and ideas taken by the North from the South in the last five centuries? And, today, what do patents mean when, for example, a plant patent of a corporation in the North is made with germplasm taken from the South, perhaps also making use of botanical 'folk' knowledge also taken from the South, and, through the brain drain, perhaps employing the best minds born, fed and educated in the South?

In any case, the main point is that the world distribution of the capacity to continue to create knowledge is becoming dangerously skewed and would deteriorate even more in a two-track world model. For the North not only concentrates all the financial, institutional and intellectual resources for research, for its development it continues to accumulate the physical and intellectual resources from the South.

True, governments in countries of the South are also at fault. Ecological destruction, cultural discrimination, government insensitivity and public disregard for local and traditional knowledge are destroying a wealth that has taken millenniums to be built up.

It is hard to exaggerate what this loss of knowledge will mean in terms of human civilization. Biologists are urgently warning, in the words of E. Wilson (1984: 121) "The worst thing that will *probably* happen—in fact is already well underway—is not energy depletion, economic collapse, conventional war, or the expansion of totalitarian governments. As terrible as these catastrophes would be for us, they can be repaired in a few generations. The one process now going on that will take millions of years to correct is loss of genetic and species diversity by the destruction of natural habitats. This is the folly our descendants are least likely to forgive us." Precisely the counterpart of this loss of genetic diversity and of species in the realm of culture is the loss of the diversity of human knowledge and of cultural traditions.

Granted, human beings also have the capacity to adapt to rapidly changing environments, and so room must be left for new knowledge. Yet this gift of adaptation is based on the one single ability that people have over plants, and still over computers, and that is the ability to learn from experience. If this experience, in varied forms of local, or traditional knowledge related to pharmacopeia, ecology, botany, zoology, agronomy, hunting, fishing and collecting, physiological and psychological therapy, and symbolic systems, is erased from the book of human history, this will mean an impoverishment, not only in absolute terms, but in possibilities to learn from experience for scientific and social advancement.

It is not only a question of folk or ethnoscience giving important leads which are then developed in sophisticated laboratories—and, by the way, patented. It is equally true that however refined the products of scientific laboratories or seminar rooms, all technology, administrative models or economic policies have to be adapted, moulded and combined with local knowledge if they are to be successful in different geo-ecological, political and social environments. Indeed, some of the worst examples of development failures can be traced to the lack of attention to local conditions.

Policy actions, very especially in the 1990s, should give attention to actively preserving this diversity of knowledge. This can be done, among other things, by using the sophistication of informatics to record, classify and disseminate such knowledge bases, and by creating awareness and supporting projects among local populations so they themselves are able to treasure and to preserve such knowledge for their own advancement.

4.5 Summing up Policy Proposals

If the working framework in which we deal with the new world trends is global and holistic, the search for shared goals becomes important. The shape that political ideologies took in the twentieth century has now become obsolete; this does not mean that the principles which guided many of them are obsolete. The best of these principles: equity, democracy, freedom of choice, human and civil rights, must be given a new shape in the forthcoming decades. *Le pari*[2]: the challenge is to shape something which is not yet born but which, in the process of being tried out, will be born.

How to induce this birth? Well, by supporting interdisciplinary research and policy groups, including both North and South participants, to rethink political, social and cultural goals and institutions on a global basis. Of course, as so many participants in the seminar proposed, new financial and economic management, as well as environmental institutions have immediate priority.

Just as important is to create greater public awareness through the mass media, both in the North and in the South as to the need for a global approach, in which shared responsibility and benefits are negotiated for a sustainable world development.

As to social sustainability concerns, research efforts should be directed at incorporating the economic costs of social malaise into economic development models. This is partly a methodological problem, whose solution should be geared, firstly, to creating hard data on such phenomena.

On the policy side, as long as economic dualism persists, the need for some form of social welfare for those excluded from the high speed economy will continue. But right beside aid or social security, policy should encourage what could be termed 'freedom of inventiveness', that is, the ability to create new means of livelihood. How? Among other things, by providing credit for micro-industries, especially for women heads of households, and for environmentally sound agricultural and fisheries micro–projects; by promoting new organizational schemes in

[2] In French, *pari* is a challenge but it also means placing a stake in something you wish for.

high technology agriculture that will foster negotiated interdependence rather than vertical control; by accelerating and broadening the dissemination of the knowledge base of new technologies and of market movements; by enhancing renewed ties of family, kinship and social solidarity which will continue to give people a sense of belonging without being oppressive to women; by supporting people's initiatives to create a new 'sense of place' in a global society.

In terms of cultural sustainability, there must be greater awareness that, ultimately, a sense of meaning and of purpose are the two rails on which any train of development must travel, and, therefore these two concerns must have a place on political agendas. Can Western culture alone provide the answers for globally shared goals? This is no longer possible in a global context. The answers must come out of a constant dialogue between cultures and through science. Such an end can be pursued through mass media programs, many kinds of research, scientific and educational programs, and, very specially, art, which explores and enhances such a dialogue.

It would be desirable that special research and policy actions be focussed on the multiethnic dimension of national societies: will negotiations among such groups be handled at the level of national governments or of macro-regions? What will the patterns of ethnic migration be from less developed countries left out of macro-regions to the latter? The knowledge base on this topic amassed by anthropological research during many decades should be tapped in this respect. Comparative studies of recent legislation and political agreements with indigenous and national minorities should be pursued.

Finally, as mentioned above, a massive effort should be made, the world over, to record local and traditional knowledge, especially in regions where both bulldozers and televisions are doing away with legacies of observation evolved during centuries. This can be done rapidly using computers and ethnographic methods. Ideally, projects by the communities themselves—the young and the old especially—to record their legacies of wisdom should be stressed.

4.6 A Time of Hard Choices

In general terms, avoiding a global and national two-tiered economy of the rich and the poor means breaking down juridical, political, scientific, and cultural barriers that polarize the use of resources. In concrete actions it means expanding educational services, preserving, acknowledging and adapting local knowledge, promoting democracy, eliminating ethnic discrimination, opening the mass media to local participation, among other things. Importantly, it does not mean giving handouts, it means rebuilding livelihoods.

As James Grant, director of UNICEF, has recently stressed, it is a time for hard choices. At this moment it is still time to choose between an apartheid pattern for global development, with all its instability, poverty and suffering, or one where the process may be an end in itself for the time being, a participatory path towards the future. This seems to me our only option: not knowing where we are going, we can still decide how to move forward.

References

Wilson, Edward O., 1984: *Biophilia* (Cambridge, Massachusetts: Harvard University Press).

Chapter 5
The Global Cube

5.1 Introduction

Is it possible for the world to become one as the UN World Commission on Environment and Development suggests in the very first sentence of its report?[1] As statesmen begin to declare that 'our country is the planet' and economic and political systems become metanational, a global perspective enters the agenda of the human race.[2] Yet it is still not clear whether such a change will lead to a more balanced world or one fraught with environmental hazards and fragmented into islands of wealth in seas of poverty both between North and South and within nations. Will it be driven by conscious, collective action of humankind, or impelled by short term, destructive processes? The way in which it goes will depend, partly, on how quickly and accurately science is able to cope with the challenge of thinking and analysing phenomena from a global perspective.

The social sciences face a fundamental challenge in studying global change. As has been pointed out, the 'sociosphere' cannot be seen with the eye when a photograph of the planet is taken from outer space.[3] As a bounded entity, we seem to exist only in our mind's eye. For this and other reasons, the impression is sometimes given that the changing textures of the geosphere and biosphere are merely natural

[1] 'The Earth is one but the world is not' is the expression used by the UN World Commission on Environment and Development at the beginning of their report to highlight the need to build sustainable forms of development for the future.

[2] This text was published first in: *Global Environmental Change. International Social Science Journal*, XLIII,4 (Blackwell Publishers. UNESCO, 1991): 599–608. This text is used by permission of UNESCO and the permission was granted on 4 July 2013 by Isabelle Nonain-Semelin on behalf of UNESCO.

[3] The French sociologist Daniel Bertaux discussed this at one of the meetings of the Human Dimensions for Global Change Committee of the International Social Science Council. I am indebted to the participants in the Committee for the discussions which gave rise to many of the thoughts put forth in this article: Dr. Harold Jacobson, co-ordinator; Daniel Bertaux, Leszek Kosinski, Robert Worcester, Kurt Pawlik, Renat Perelet, Takashi Fujii, Ashish Bose and Martin Price.

L. Arizpe, *Lourdes Arizpe*, SpringerBriefs on Pioneers in Science and Practice 10,
DOI: 10.1007/978-3-319-01896-6_5, © The Author(s) 2014

phenomena when, to a large extent, they are subject to a human-driven process, one deeply involved with the human use of the resources of the planet.

At present, the physical and natural sciences are advancing rapidly in the creation of knowledge as to how, where and at what rates such resources are being used. In contrast, the social sciences ought to be asking why such resources are being used in the way they are, that is, to what human purpose and—most importantly for seeking solutions for a sustainable future—who is making use of them. There are many instances where sound policy recommendations for sustainable development have become ensnared in the web of interests or disinterest, of diverse institutions and social groups.

This new field of research is being widely referred to as 'global change'. Martin Price rightly argues that this has been defined predominantly from a geocentric and biocentric perspective so that advances must be made in developing a sociocentric perspective (Price 1989). A more balanced perspective, however, will take time to develop, primarily for epistemological reasons, but also because the term is being utilized indistinctly to refer to new kinds of *phenomena* and also to a *new perspective on* world reality. Occasionally, it is also applied to designate a 'stage' in world civilization and to call for a new 'global' science.

This chapter discusses how the 'global' is being interpreted and therefore seen in research programmes in the natural and the social sciences and focuses on basic theoretical and modelling problems that must be overcome in order to establish an adequate basis for a comprehensive research programme on the plurality of phenomena that are now being included in global environmental change.

5.2 A Joint Agenda with the Natural Sciences

Natural scientists have already identified the most urgent global environmental problems (among them climate change, ozone depletion, loss of biodiversity) in the International Geosphere–Biosphere Programme: a Study of Global Change (ICSU 1989; Jacobson 1990). A first step towards collaboration between the natural and the social sciences has been taken in *A Framework for Research on the Human Dimensions of Global Environmental Change* of the Standing Committee on the Human Dimensions of Global Change, of the International Social Science Council (Jacobson/Price 1990).

However, given the complexity of human actions associated with the multiple processes of global environmental change, it is necessary to develop a long-term programme of theories and methods within the social sciences to cope with what now seem historically unprecedented situations. It is not enough to draw up an *agenda* of topics for research in the social sciences related to global change; what is needed is a new global *perspective* that will allow us to dissect existing reality in a way different from that which has informed social science research for the past century.

Attempts at building such a new global perspective are already under way. Burton and Timmerman (1989: 302) draw our attention to 'a new paradigm' that is emerging, based on what they call "the evolution of 'complex systems'". However, building a new paradigm would entail not only creating a new theoretical programme for research but also reviewing some of the basic epistemological assumptions of regular social science. As there are intrinsically associated with fundamental philosophical tenets, it means that new philosophical and ethical norms should form part of the programme for creating new knowledge. These must be developed to cope with new areas of the human experiences, e.g. genetic engineering, as well as to the new limits to human drives, e.g. the depletion of non-renewable resources.

One example of a promising avenue for thinking about global reality is the theoretical scheme being developed by Gilberto Gallopín and the researchers at the Fundación Bariloche. They propose the use of the concept of 'global impoverishment' to encompass *both ecological and economic* impoverishment as the central process of global change (Gallopín/Gutman/Maletta 1989; Gutman 1988; Leff 1986).

What are the criteria, then, with which we may begin to define the 'global' in the social sciences?

5.3 The Definition of Global Change in the Natural Sciences

Menon (1989: 60), President of the International Council of Scientific Unions has defined global environmental change as follows:

> Humankind has no doubt been altering the environment in the process of living and development for at least 2 million years, but during most of this time, human influence on the environment has been local in scale and small in magnitude. It is only over the last half a century that humankind has developed the ability to alter the environment on a global scale, and not just in terms of local effects such as due to pollution.

The criterion for defining the 'global' here is that of *scale*. Additionally, Menon also mentions, in his description of the key objective of the International Geosphere Biosphere Programme (IGBP), that global change involves "interactive physical, chemical, and biological processes that regulate the earth system, the unique environment that it provides for life, the changes that are occurring in this system, and the manner in which they are influenced by human actions" (Menon 1989: 60). The second criterion used to define it, then, is the interaction between the different processes that are seen as its components.

If the main criteria to define global phenomena in the natural sciences are scale and interaction, it would be easy to create mirror image of such phenomena in the social sciences, i.e. they would be equivalent to the 'internationalization of the market economy', the 'globalization' of telecommunications, transport, and the mass-media and so on. Understanding such phenomena would imply merely adding a new level of magnitude to our models and establishing the linkages between the different levels.

But changes which are *seen* as global, as far as human groups are concerned, are associated not only with change in the speed, density and scale of interactions, but also with change in the structure and complexity of such interactions. As Miller (1989: 87) has stated, to explain global environmental change it is necessary to examine the *direct human actions* which influence it, as well as the *indirect human actions* that set in motion complex chains of events which also affect the environment. It is the study of such indirect human actions embedded in the dynamics of change of political and economic systems that require the development of theoretically autonomous social theories to explain them, independently from environmental concerns, yet linked to global problems.

Thus, research on global change must go further than simply measuring what is already *seen,* and face the more arduous task of asking how global change may be *interpreted.*

5.4 The Global Cube

To make a statement about our perception of reality, Wittgenstein used the example of a three-dimensional drawing of 12 lines that could be interpreted, and thus seen, alternatively as a cube, a wire structure, six boards and so on. Paraphrasing him, one could say that 'we *interpret* the global and *see* it as we *interpret* it' (Wittgenstein 1958: 200). How do the social sciences interpret the world and, therefore, the global? Inevitably, on the basis of a world view inherited from nineteenth century epistemology.[4]

Actually, one could question the validity of applying this to all the social sciences, but a case can be made for this by arguing, firstly, that all of them are based on the assumptions of scientific rationalism, and, secondly, that all of them, including economics, are having difficulty in adjusting their lenses to contemporary phenomena. However, because it is highly complex to generalize for all the social sciences, the following remarks are necessarily schematic, for the purpose only of opening a debate.

5.5 The Interpretation of the Global in the Social Sciences

One possible interpretation of the term 'global' which would make it a withered old claim in the social sciences is to consider it synonymous with 'universal'. The main objective of positivistic social science has been to arrive at universal laws of

[4] I am indebted to Benjamin Mayer for the discussion of the next three sections.

human experience, an intellectual endeavour going all the way back to the Greeks. But the introspection of the generation of the 1960 s in Western social science, and the critique made by Third World scholars against 'ethnocentrism' in the 1970s showed how the social sciences had fallen short of their goal of universalism.

Furthermore, not only has the current failure of universalism been diagnosed; more radically, the *impossibility* of an absolute universalism has been theorized, notably in critical theory. In other words, the current failure of universalism is due to the limits of our current working tools: it is an *essential* failure or impossibility.

A relativistic, as opposed to a universalistic perspective follows from such a research programme. Against this 'revival of relativism', Archer (1987: 249) argues that 'if we cannot ascribe beliefs, the end result is that sociology has *no* role to play in explaining action. This must be handed over to behaviourism, materialism or indeed biology—in short, anything which excludes reference to the determinate beliefs of human subjects. Geertz (1984), from the viewpoint of anthropology, offered an Anti–Anti-Relativism argument stating that we are adrift like the Polynesians on the high seas and we must learn to live with intellectual uncertainty.

Critical theory goes even further by questioning the very language being used to think about reality. The conceptual framework of science's approach to 'the global' would require detailed linguistic and philosophical examination. In this sense, the social sciences are considered exercises in exegesis, dealing with interpretations of interpretations of interpretations. Thus, using the term 'global' to refer to the classical domain of universalism becomes highly problematic.

Another theoretical difficulty in dealing with global change stems from methodological individualism as applied to social analysis. The basic assumption of models using this methodological scheme emphasizes that all social phenomena are to be analysed in terms of individuals' beliefs, property and actions. One such model which is currently highly influential is the rational choice model.

It would seem, though, that in studying global phenomena, such a scheme makes it enormously difficult to bridge the gap between the choices individuals make at the very micro level, and the global effects of myriad such choices. Also, if we are facing historical thresholds related to global change, one of Przeworski's (1987) objections to this type of analysis holds true; that is, that individual preferences are neither universal nor stable and depend, instead, on conditions and thus change through-out history. Another objection he makes also applies—that, "under certain conditions, a rational action is not possible even if individuals are 'rational"' (1987: 104). In fact, what is happening today in global environmental change seems to epitomize the clash between the rational choice made by individuals wanting to prevent further environmental depletion and the irrationality of economic and political systems that cannot prevent unsustainable environmental actions.

An additional, more general problem in developing a global perspective in the social sciences is the inherent difficulty most of them have always had in explaining social change, especially in certain theories such as functionalism, and even Marxism. As Przeworski (1987: 104) has emphasized, "Marxism is a theory of history with no theory as to the actions of people who make that history."

This is related not only to the assumptions of such theories but stems from a much more widespread bias in classical science. Now that the study of non-linear, irreversible processes is gaining ground in the physical sciences, it shows how intent classical science was on focusing stable, linear and reversible phenomena. A bias towards stability has been present in the very *choice* of problems selected for research, e.g. atmospheric science can fully explain how atmospheric gases are constituted and interact, but they cannot predict tomorrow's weather.

Because the social sciences accepted the norms of the scientific method as developed in the natural sciences, this same implicit bias towards choice of stable phenomena as objects of study was also transferred to the former. This can be illustrated by the fact that, until the 1970s, most sociological and anthropological studies concentrated on communities or societies; only in the last few decades have social movements, migration or changing family patterns been dealt with. Moreover, what had been considered 'deviant' behaviour in the 1950s became 'normal' in the 1980s.

Global change, related to the environment or to other social phenomena, is essentially a study of movement, of crossing historical thresholds into new domains. This transition is difficult to grasp to analyse with the classical tool-box of the social sciences. But since we social scientists are also involved in creating these new domains, we must exercise our ability not only to analyse what is happening, but to help evolve the new philosophies, new ethics, new behaviour and new social pacts needed to make this new world sustainable.

However, present social science, with its empiricist emphasis, seems to be caught in a double bind: the more rigorous it becomes, the less it is open to creativity. At the time that the social sciences were developing in the second half of the nineteenth century, the multiple debates and publications of a philosophical and speculative nature on society and on the individual provided a rich humus of ideas out of which grew the basic hypotheses of the social science. Today, because there tends to be an exclusive emphasis on empirical observation and measurement in the social sciences, philosophical thinking and intellectual adventure have been left out of major debates. For example, political science, according to Trent (1990), has been hindered by a reductionist view of its domain, while political philosophy has been unwisely neglected.

To put it even more strongly, because global phenomena are unprecedented in the human experience, we not only lack the methods to apprehend them, we lack the basic categories and ideas with which to *think* about them. Thus, we have few hypotheses, in the traditional body of theoretical thinking in the social sciences, to *interpret* such phenomena. This may help explain why global phenomena seem to be so invisible from the standpoint of the social sciences. Since our *interpretation* of the world in the social sciences is still so constrained by nineteenth century mental maps of one great divide, the West and the Other Peoples, each atomized into hundreds of nations or 'tribes', it is hardly surprising that we cannot *see* global phenomena occurring in our everyday experience.

5.6 The Anthropological Experience

In spite of the fact that anthropology was created in the 19th century to deal with the 'Other', with the orphan half of the political dichotomies of civilized/primitive, history/no history, we/they, and stretching the *intended* assumptions of the discipline a little, that is, the aim of explaining human nature in all societies, it can be said that anthropology was intended to be, in fact, *a global social science*. This is meant in the sense that its field of study encompasses the whole of the human past, on a time range of millennia, and that a major thrust of its research deals with the interaction between human societies and the natural environment.

It is worth noting that this knowledge has shown, among other things that this is by no means the first time in history that ecological factors have made the course of human civilizations swerve in unexpected directions. To illustrate this, one can turn to Egyptian civilization, whose rise has been associated with the desertification of nearby regions that drew agricultural populations into the fertile Nile valley and led to the emergence of a highly advanced civilization (Manzanilla 1989). On the other hand, many examples exist of highly developed central states which most probably fell because of political and social strife resulting from the depletion of surrounding environmental resources, as has been argued in the case of the Maya city of Copan (Abrams/Rue 1988).

Anthropologists also produced a world map of all human cultures and then tried to generalize through cross-cultural analysis, as in the Human Relations Area Files project initiated by Murdock (1967). Did the sum of all these discrete cultural entities create a 'global reality'? Hardly. However, there are many similarities in the problems of theory and method posed by the HRAF project and those being discussed at present in dealing with data and models in global environmental change. Only two of the most important will be mentioned here.

Foremost is the problem of establishing the heuristic boundaries of the units of analysis. In cultural analysis it means deciding whether one culture should be considered a single unit in spite of internal diversity or whether its subcultures should be considered discrete entities. This methodological partitioning decisively alters the statistical distribution of cultural patterns, and, hence, the generalizations that could be made as to the nature of given cultural traits or institutions around the world. Thus, such boundaries pose not only technical problems of statistical measurement but theoretical problems related to the way in which the relationship between the different cultures is defined.

To illustrate how this problem is related to studying global change, consider the following example. In studies of deforestation, if the unit of analysis is defined as a 'human population' and this is related to the amount of land or natural resources available to this population, there is an inherent bias towards considering that every individual has equal access to such resources. Instead, if the units of analysis are defined as groups which have a significant relationship to the decision-making institutions which control access to such resources, the interaction of such groups with environmental depletion is inherently assumed to be one not exclusively of demographic numbers but one associated with the access that each group has to the use of resources.

A second problem that was also present in mapping out the cultures of the world was that once the discrete cultural units were identified, it became difficult to explain the dynamics of the relationships between them.

In studying the human dimensions of global environmental change the same kinds of problems are present. How can significant social groupings be related to the 500 km^2 units of Geographical Information Systems? Are political or administrative boundaries—national, state, provincial, municipal—useful in defining the units of analysis for studying environmental depletion? Should ethnic boundaries be considered important in analysing environmental change?

5.7 The Clamour for Interdisciplinarity

The clamour for interdisciplinarity in social analysis can be heard everywhere, and particularly in the study of global change (Dogan/Pahre 1989). The more disciplinary domains are analysed, the more evident it becomes that national and institutional boundaries had a major role to play in defining such domains. It is worth noting that since the nineteenth century, geo-political boundaries have been used conveniently as a methodological tool in delimiting the objects of study in the social sciences.

The first political boundary, as mentioned above, was that between Western and non-Western peoples which led to the creation of a special social science, anthropology, to study the latter. Other social sciences were developed for the study of Western national societies. For example, political science is founded on the concept of national sovereignty; economics has dealt mainly with national accounts and national markets; sociology is focused on social institutions of Western societies; psychology—and the concepts of mental health and illness—were developed and applied mainly to European mentality and behaviour. It has repeatedly been pointed out that with the expansion of regional economic and political blocs, of the market economy, and of ethnic strife, nation–states are weakening and so are the theoretical constructs based on this central institution.

As has been pointed out by Dogan and Pahre, who suggest that the term interdisciplinary be substituted for multidisciplinary, hybrid fields of study in this field are expanding rapidly. Examples are, women's studies which utilize theoretical and methodological tools taken from many disciplines; biological anthropology, artificial intelligence and communications theory.

According to these authors, the hybrid fields are advancing in the social sciences through a process of 'fragmentation-into-hybridization' (Dogan/Pahre 1989: 457). The more the research field of global environmental change expands, the more hybrid fields will develop, especially along the borders of the natural and the social sciences, for example, human ecology or environmental economics.

5.8 Microsocial Models in Global Contexts

What kinds of micro-social models can we begin to build in order to study environmental change? In what follows, a few of the major theoretical problems in setting up such models are discussed, in relation to the decision-making process in a case study of deforestation.[5]

The first problem is how, in a global context, to define the process under study. In his pioneering work on modelling for global environmental change, Rolando/ García (1986) proposed a three-tiered structure linking micro to macro processes. To capture adequately the dynamics of such processes, he stresses that the property of the system is not found in its components but in the relationships between its components. Thus, it is necessary to use process oriented models. One such model deals with the series of decisions that lead up to a major event or phenomenon.

It is well known that social scientists working at the micro level of human behaviour have always faced the intractable problem of explaining how individual patterns create statistical frequency distributions around a mode. The most readily observable phenomenon to explain such distributions is that of the choices made by individuals. However, it is one thing to map out individual choices, and another to try to explain the constraints and incentives that channel individual choices. The first would explain what individuals tend to do, the second, how social patterns are created.

Choice, however, may be theoretically defined in terms of individuals' *motivations,* as in the rational choice scheme, or it can be posited as an *action which generates social structures.* Choices are indeed made by individuals but such choices are embedded in *exchanges.* When it is stated, for example, that a market economy allows individuals to choose freely, the fact is overlooked that such choice implies a *transaction with another individual or entity,* in which both may gain or lose, or one gain more than the other. The consequence of such transactions pertains not only to the first individual's well-being, but, more importantly, to the well-being of society as a whole. Thus, in what follows, choice is defined not in terms of an individual's psychological motivation, but of options taken by individuals or groups within a given range of alternatives.

It is important to stress that the individualistic bias of choice models is well suites to Western industrial societies but not to agrarian or non-Western societies which have strong communal or corporatist traditions. In the latter, social networks and corporate groups must be taken into account in defining the units of analysis in research.

Thus it seems that the explanatory value of choice analysis comes not from ascertaining that individuals may have a maximising attitude towards the economy, the environment or life generally, but from the range of options provided

---

[5] This study on 'Perceptions of social and environmental change in the Lacandon rainforest in Southwest Mexico', directed by Lourdes Arizpe, is funded by the National Autonomous University of Mexico (UNAM).

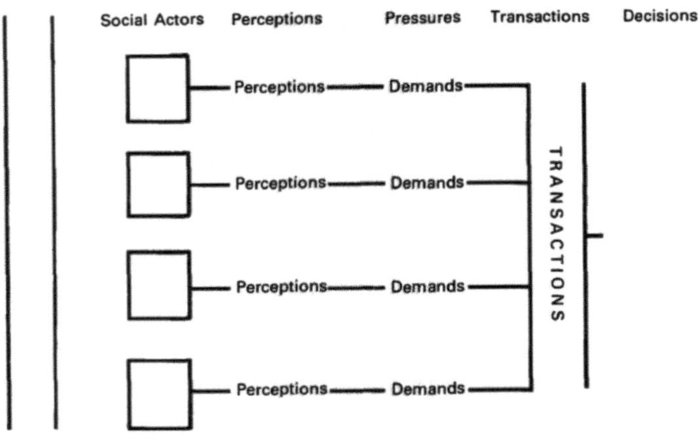

Fig. 5.1 Methodological tools used for case study. *Source* The author

by the political, economic and social systems and the social pressures that trigger choices.

Another important issue in linking the micro to the macro level in global models is that the forces generating environmental depletion cannot be understood unless the time factor is incorporated into the models. Pressures that alter the subsystems at different levels of magnitude arise at different times and cause chain reactions at other levels of magnitude. Only a diachronic approach can help explain the dynamics of these reactions.

The model being used in a case study of deforestation in Mexico recognizes three levels—local, regional/national and international/global—at which processes have a significant systemic behaviour. The following methodological tools are used (see Fig. 5.1):

1. *Events.* A diachronic approach always has the problem of how to carve up a continuous sequence of phenomena. Using an approach similar to that developed by Kates (1979) to study climate change, *events* are identified as those decisions or social movements which decisively alter the course of a process, in this case, the acceleration of the rate of deforestation in Lacandona.

2. *Initial conditions.* The set of initial conditions may be defined as those existing immediately prior to a major event that caused a sufficiently significant alteration of these conditions that their rippling effect can still be discerned in present conditions. Conditions can be distinguished from events in that the former have a longer, more durable time span than the latter, and usually pertain to the levels of higher magnitude of phenomena.

The usefulness of such a model is the possibility of going backwards towards a previous bifurcation to explain generative factors in historical time, which actually move between different levels of magnitude.

3. *Rationalities and choices.* In a most illuminating article, Schmink (1989) explained the rationalities of different social groups involved in Amazon deforestation. I have taken the same scheme, of identifying social groups having a significant impact on deforestation and explaining the rationality of their behaviour. However, instead of looking at it as one process made up of different rationalities, I have divided it up into events, in which significant social groups may play a different part in a given event. The degree of pressure exerted by each social group at that precise point in time is measured and brought to bear in the transactions which occur between the groups. And the outcome of the event is evaluated relative to the range of options which existed as a function of the total transactional process.

In defining this range of options and in the balance that each group makes of its 'gains' or 'losses', perceptions play a major role: perceptions not as psychological states but as expressions of cultural values in a context of accelerated social change. This fundamental stratum of cultural values is the basis of assessments of options and choices made (Clark 1989). Thus, the perception of reciprocity, gain or loss is also incorporated into the model being used.

It seems to me that this kind of model may open a path to obtaining better answers to questions of why some social groups act in apparently irrational ways that endanger the environment and planetary survival. And it may be especially useful to make policy recommendations.

The dynamics of rapidly changing groups of social actors can be analysed. Differing perceptions of what diverse groups consider 'gains' or 'losses' can be assessed. And the total systemic process of exchange between groups can be explained for each event, so that policy can be applied precisely at the point where the offer of other options to specific groups can deter behaviour deleterious to the environment.

5.9 Conclusion

Close collaboration with physical and natural scientists is indeed necessary in studying the most urgent problems of global environmental change, but this is only the first step in developing a social science perspective of global change. But to *see* the social phenomena related to global change it is not enough to find the proper scale, measurement tools or inter-linkages between the global and the micro levels of analysis. Instead, we need to analyse the *interpretation* of global reality which is built into the theoretical assumptions that created disciplinary boundaries and mainstream categories of analysis. This interpretation was established for a world divided into Western and non-Western peoples, fragmented into nation–states and regulated by relatively stable institutional domains.

Part of the problem is that many of the most influential epistemological and theoretical schemes, i.e. methodological individualism, logical positivism, as well as relativism and deconstructionism create methodological barriers to building a global perspective of social phenomena.

There are, however, new theoretical schemes developing, and multidisciplinary fields emerging which are laying the groundwork for methods with which to approach global environmental change. The urgency in advancing our knowledge of this new reality cannot be under-stated. The use of the planet's resources, the imbalances of over-consumption and under-consumption between social groups and countries, and the pattern of our own lives, all require that a new interpretation be created which will make us *see* them as parts of one global reality. Data and models to study global change are the tools with which we can experiment and build such a new interpretation and a new world.

References

Abrams, Elliot M.; Rue, David J., 1988: "The Causes and Consequences of Deforestation among the Prehistoric Maya", in: *Human Ecology*, 16: 377–395.

Archer, Margaret, 1987: "Resisting the Revival of Relativism", in: *International Sociology*, 2,3 (September): 235–250.

Burton, I.; Timmerman, P., 1989: "Human Dimensions of Global Change: a Review of Responsibilities and Opportunities", in: *International Social Science Journal*, 121 (August): 279–313.

Clark, William C., 1989: "The Human Ecology of Global Change", in: *International Social Science Journal*, 121: 315–345.

Dogan, Mattei; Pahre, Robert, 1989: "Hybrid Fields in the Social Sciences", in: *International Social Science Journal*, 121 (August): 457–470.

Gallopín, G.; Gutman, P.; Maletta, H., 1989: "Global Impoverishment, Sustainable Development and the Environment", A Report to IRDC on the Project 'Global Impoverishment and Sustainable Development (Mimeo).

Geertz, Clifford, 1984: "Anti–Anti-Relativism", in: *American Anthropologist*, 86,2: 263–278.

Gutman, Pablo, 1988: *Desarrollo Rural y Medio Ambiente en América Latina* (Buenos Aires: CEAL).

ICSU (International Council of Scientific Unions), 1989: *The International Geosphere-Biosphere Programme: A Study of Global Change* (Paris: ICSU).

Jacobson, Harold K., 1990: "Note on the Human Dimensions for Global Change", in: *HDGEC Internal Document*, No. 2.

Jacobson, Harold K.; Price, Martin F., 1990: *A Framework for Research on the Human Dimensions of Global Environmental Change* (Paris: HCGEC-ISSC-UNESCO).

Kates, Robert, 1979: *Climate Change Assessment* (Chicago: Chicago University Press).

Leff, Enrique (ed.): 1986: *Los Problemas del Conocimiento y la Perspectiva Ambiental del Desarrollo* (México City: Siglo XXI).

Manzanilla, Linda, 1989: *La Formación del Estado en Babilonia* (México City: UNAM).

Menon, M.G.K., 1989: "Opening Address", in: *Global Change*, Report 71.

Miller, Roberta, 1989: "Human Dimensions of Global Environmental Change", in: De Fries, R.S.; Malone, T.C. (Eds.): *Global Change and our Common Future* (Washington: National Academy Press): 84–89.

Murdock, George, 1967: *Ethnographic Atlas* (Pittsburgh: University of Pittsburgh Press).

Price, Martin, 1989: "Global Change: Defining the Ill-defined", in: *Environment*, 31, 8 (October): 18–20 42–44.

Przeworski, Adam, 1987: "Marxismo y elección racional", in: *Zona Abierta*, 45 (October-December): 97–136.

Rolando Víctor, García Boutigue, 1986: "Conceptos básicos para el estudio de sistemas complejos" in: Leff, Enrique (Ed.): 1986: *Los Problemas del Conocimiento y la Perspectiva Ambiental del Desarrollo* (Mxico City: Siglo XXI): 381–409.
Schmink, Marianne, 1989: "The Rationality of the Destruction of Tropical Forests", University of Florida at Gainesville (Mimeo).
Trent, John, 1990: "Though on Political Thought: An Introduction", in: *International Political Science Review*, 11: 5–23.
Wittgenstein, Ludwig, 1958: *Philosophical* Investigations (New York: Harper and Row).

Chapter 6
The Culture of the Ax, the Machete, and the Sling

The previous chapter (Arizpe/Paz/Velazquez 1996, Chap. 3) presented the Lacandona forest inhabitants' concepts of nature. This chapter describes the ethnography of their perceptions on recent changes in the natural environment, particularly the deforestation, of the Lacandona rain forest.[1]

Two different methodological operations will be involved. Ethnography takes us back to the classic method of anthropology, in which the norms obtained from informants were recorded within a well defined semantic field. We used this method in fieldwork fully aware of its inherent trend toward "the nostalgia for the origin, the archaic and natural innocence, the purity of the presence and the presence itself of the word", as Derrida very poetically put it (1978: 95). We followed this course, admitting, following Lévi-Strauss (1973), that its lack of acceptable historical significance is offset by the value justifying it as an instrument of method.

We found, however, that norms relating to deforestation did not exist but were, rather, in the process of being created. In a sense, then, there was no normative centre in people's utterances. Rather, their views ran along the borders of emerging lines of discourse.

Therefore, our second methodological operation was to map out the conceptual territory of the emerging perceptions of environmental change in the region. The first task was to find out what boundaries were established in our informants' discourse related to such change.

---

[1] Arizpe, Lourdes; Paz, Fernanda; Velazquez, Margarita, 1996: *Culture and Global Change: Social Perceptions of Deforestation in the Lacandona Rain Forest in Mexico* (Ann Arbor: The University of Michigan Press). The permission to republish this chapter was granted on 9 July 2013 by Mr. Aaron Mc Collough, University of Michigan Press. This book is available at: <http://www.press.umich.edu/14545/culture_and_global_change>.

L. Arizpe, *Lourdes Arizpe*, SpringerBriefs on Pioneers in Science and Practice 10,
DOI: 10.1007/978-3-319-01896-6_6, © The Author(s) 2014

6.1 External Pressures, Internal Perceptions

Environmental change, particularly, deforestation, soil erosion, water pollution, and the loss of rain forest flora and fauna have been going on in the Lacandona rain forest for decades, yet until recently they had not been converted into an issue of public life.

Local inhabitants undoubtedly perceived these changes in the sense of "receiving the impression of an object through the senses" (RAE 1984: 1041). Likewise, some of them no doubt understood these phenomena, in the sense of "discovering the substance, quality and relation between things through the exercise of the intellectual faculties" (RAE 1984: 361). Such perceptions and understanding acquired a social and political significance, however, which turned them into public issues because of information and pressures coming from outside the region.

From the 1960s onward, as we have seen, the government, development agencies, and state entrepreneurs all agreed that the forest was 'unproductive', a word still used by many in the region, especially large-scale cattle ranchers. They insist that productivity can only be increased through agriculture and cattle raising, which means doing away with the rain forest.

At the beginning of the 1970s a few environmental groups began to emerge in favour of conservation of the rain forest. They were lead by Gertrudis Duby and Manuel Alvarez del Toro, and their efforts were successful in creating a broad coalition of environmental groups in Chiapas linked to other national and international nongovernmental organizations.

By the end of the 1980s the information conveyed from the outside world to Chiapas by the mass media, the shift in government policies toward the rain forest, the appearance of international agencies interested in conservation, and the multiplication of environmental groups all pressed for conservation. Consequently, deforestation, together with other environmental issues, came to the fore in public discussions in the whole of Chiapas.

To find out how local people were giving semantic form to this change in their perceptions, we conducted a survey of seven communities: two ejido communities settled in the 1970s (Pico de Oro and Reforma); two ejido communities settled in the 1980s (La Victoria and Nuevo Chihuahua), the four located in Marqués de Comillas; two rural communities settled in the 1960s near Palenque (Lacandón and La Unión); and Palenque itself, with specific focus on three sub–communities: one low-income urban group; one high-income urban group; and one cattle-ranching group (more information is provided in the *Appendix* in Arizpe/Paz/Velazquez 1996: 103).

6.2 What Do You Think is the World's Greatest Danger?

In the answers to the open-ended question 'What do you think is the world's greatest danger?' we found at one end of the scale those who consider life itself as dangerous, so that everything is a threat. At the other end were the optimists, who declared that "everything is fine; there's no problem".[2] Between these two extremes there were seven main types of concerns, regarding: war, 23.8 %[3]; poverty, 13.7 %; pollution, 8.3 %; attitudes, 8.3 %; illness, 7.9 %; deforestation, 6.7 %; environmental degradation, 4.4 %; divine punishment, 4.4 %, overpopulation, 3.2 %; and others, 3.5 %.[4]

One response bears mentioning, which, although unique in the survey, was sometimes mentioned by other informants during field-work. This is the view, given by a farmer from Pico de Oro, that the forest itself is dangerous: "Here, we're surrounded by dangers, animals, snakes, falling sticks, and crocodiles." This is, of course, true and worth pointing out to offset the sometimes idyllic and inoffensive image of the rain forest created by the mass media. The inhabitants of the forest have to fear these dangers, and for them conserving the forest would mean preserving these dangers. By way of example, in 1991, the twelve-year-old eldest son of a former *ejido* commissioner from Pico de Oro died from a snakebite. Experiences such as this necessarily inform the concepts local people have of the forest and, incidentally, of the number of children they wish to have.

Significantly, only 6.7 % of those surveyed regarded deforestation as the most imminent danger, and even this could have been a response that they thought we wanted to hear.

The differences in the responses to our question were analysed by region, community, occupation, sex/gender, ethnic group, and religion. Table 6.1 shows the differences by community, with the highest income groups and cattle ranchers from Palenque giving similar answers, related mainly with pollution and, to a lesser extent, with environmental degradation, overpopulation, and attitudes.

In contrast, the low-income groups from Palenque and the farming communities near Palenque mostly mentioned war, poverty, and attitudes as the greatest dangers facing the world. There is greater variation among rain forest communities, which point to war, deforestation, illness, and divine punishment as pressing dangers.

The contrast also holds if the first four priorities for the answers are compared (see Table 6.2). They clearly tend to vary according to dyads: high-income residents and cattle ranchers in Palenque; low-income residents in Palenque, and Lacandón-La Unión; and, to a lesser extent, the Pico de Oro-Reforma and La Victoria-Nuevo Chihuahua groups.

---

2 For information on the survey, see Appendix in: Arizpe/Paz/Velazquez (1996): 103.

3 'War' is an important response because the survey was carried out at the time of the Persian Gulf War, a fact that should be taken into account.

4 'Other' included various responses, such as lack of technology, natural disasters, the forest ban, and the forest's natural dangers.

Table 6.1 Percentage distribution of answers on the greatest danger in the World today, by community

Answers	Total	PR (1)	VC (2)	LU (3)	PL (4)	PH (5)	CR (6)
Poverty	13.7	15.6	9.4	15.6	18.8	6.3	16.7
Deforestation	6.7	17.7	8.3	–	4.2	2.1	2.1
War	23.8	17.7	17.7	35.4	31.3	31.3	10.4
Pollution	8.3	1.0	2.1	5.2	4.2	25.0	29.2
Illness environmental	7.9	4.2	8.3	13.5	14.6	–	4.2
Degradation	4.4	3.1	6.3	2.1	4.2	8.3	4.2
Overpopulation divine	3.2	4.2	1.0	–	–	8.3	10.4
Punishment	4.4	2.1	11.5	4.2	2.1	–	2.1
Other*	3.5	6.3	5.2	2.1	–	–	4.2
Don't know	15.7	20.8	25.0	14.6	10.4	6.3	4.2
Total	100.0	100.0	100.0	100.0	100.0	100.0	100.0

Key 1 Pico de Oro and Reforma Agraria, *2* La Victoria and Nuevo Chihuahua, *3* Lacandón and La Unión, *4* Palenque low income, *5* Palenque high income, *6* Palenque cattle raisers
*Includes lack of technology, natural disasters, the Mexican government's ban on tree felling, and natural dangers of the forest

Table 6.2 Priorities in the perception of the greatest danger in the World today by community

Community			Perception	
PR (1)	Deforestation	Poverty	Attitudes	Other
VC (2)	Divine punishment	Poverty	Deforestation	Illness
LU (3)	Poverty	Illness	Attitudes	Pollution
PL (4)	Poverty	Illness	Attitudes	Deforestation
PH (5)	Pollution	Attitudes	Environmental Degradation	Overpopulation
CR (6)	Pollution	Poverty	Attitudes	Overpopulation

Key 1 Pico de Oro and Reforma Agraria, *2* La Victoria and Nuevo Chihuahua, *3* Lacandón and La Unión, *4* Palenque low income, *5* Palenque high income, *6* Palenque cattle raisers

Although they share a common concern about poverty, those in rain forest communities emphasize deforestation and divine punishment, whereas in Palenque and the surrounding areas there is a greater emphasis on pollution and illness. Most of these differences, however, are gender based, since it is the men in the forest who are most concerned about deforestation and the women in Palenque who refer to pollution, as shown in Table 6.3. There is a simple explanation for this. It is the women who look after sick family members, particularly children, and the living conditions of low-income groups in the deforested area of Palenque tend to foster illness due to unsanitary conditions and a lack of hygiene, which are also results of pollution.

In comparison, there is little difference in the answers given by Indians and non-Indians. The survey showed that Indians are not worried about overpopulation and are less concerned than non-Indians about poverty, pollution, illness, and environmental degradation. Conversely, they are more concerned about attitudes.

Table 6.3 Percentage distribution of perceptions of the greatest danger in the World today by gender

Answers		Gender	
	Total	Men	Women
Poverty	13.7	14.8	12.5
Deforestation	6.7	8.3	5.1
War[a]	23.8	25.9	21.8
Pollution	8.3	7.9	8.8
Illness	7.9	4.6	11.1
Environmental degradation	4.4	5.1	3.7
Overpopulation	3.2	3.7	2.8
Attitudes	8.3	6.9	9.7
Divine punishment	4.4	3.7	5.1
Other[b]	3.5	3.7	3.2
Don't know	15.7	15.3	16.2
Total	100.0	100.0	100.0

[a] The high number of responses regarding war as the greatest danger in the world was influenced by the fact that the survey was taken during December 1990 and January 1991, when the war in the Middle East was imminent. This shows the impact of the mass media on remote communities
[b] Includes the lack of technology, natural disasters, the Mexican government's ban on tree felling, and the forest's natural disasters

In Their Own Words

War

Interestingly, at the time the survey was conducted we interpreted the response 'War' as a reaction to the Gulf War, but the Zapatista uprising of 1994 would now indicate that perhaps some respondents, especially women, who had knowledge of the training camps of the Zapatistas, may have been expressing their misgivings at the coming events.

One answer that exemplifies the new perception of globalization was that given by a young mother in La Unión who, having seen the signals of war on the other side of the planet, went on to say, "Now diseases come through the air, like it says in the leaflets; they come and get our children, and now we're going to get all that gunpowder."

Poverty

Most answers, in both the survey and the interviews, talked of the fear of poverty in the following terms: "The crisis is really bad. You can't support a family anymore, as far as I can tell"; "Now there's no harvest and nothing to eat." A Chol peasant farmer stated, "Life gets more difficult every day because there are a lot of peasants out of work." The words of Raúl Márquez, from the San Manuel *ejido* near Palenque, summarize the pessimistic view of present conditions and help explain how the Zapatistas gathered grassroots support:

Things are going from bad to worse. Everything's going to go up, and what we earn isn't enough. And the land's stopped producing. It has to be fertilized.

We don't get any help, from the government or anyone. We're all in a bad way, and the president doesn't give us any help. He helps the cattle ranchers, though. It's worse for people with children.

Poverty was emphasized even more in those communities that depend on wage labour: Lacandón, La Unión, and the Palenque low-income group. It is second in importance in all the communities in Marqués de Comillas, which are at risk from fluctuations in harvests and agricultural prices.

Pollution

When we asked further questions from those who had given the answer 'pollution' in the survey in Palenque, it became clear that most of them meant air pollution, which, in fact, is not an environmental problem in Palenque. Further inquiries cleared up the mystery: they were simply repeating what they had seen on Mexican television, in which constant mention is made of air pollution in Mexico City. This corroborates the idea that 'people do not live where they live' but, rather, where they are made to live by the mass media.

Other answers referred to pollution from trash and open sewers, which, together with water scarcity, is one of Palenque's visible problems. Trash is thrown into riverbeds and vacant lots, newly arrived families settle in these lots, and the pollution translates into illness, particularly among children.

Deforestation

Deforestation was a priority only in Pico de Oro and Reforma Agraria, both old settlements in the rain forest. This was probably due to the intense development projects which the PASECOP, a government program carried out in these communities at the end of the 1980s. Farmers' meetings and workshops had been held to discuss sustainable agriculture and rain forest conservation programs.

In both Pico de Oro and Reforma Agraria people generally were environmentally aware: "The greatest danger is the ending of the forest, because it acts as a lung for the world, and the oxygen would all get used up"; "We have to stop destroying the forest, it was a good thing the government put an end to the destruction." To halt deforestation, one farmer suggested; "First of all, you have to make people aware. For me the greatest danger is, if we farmers keep cutting down the forest, we'll use it all up."

Interestingly, some of them have environmental knowledge from previous situations. In Reforma Agraria respondents pointed out the danger of "destroying the forest, because it won't rain anymore. In Oaxaca there was just forest. They destroyed it, the river dried up, there were no more fish, and the land was no good anymore." They fear a similar future in Marqués de Comillas. Many have already made the link between deforestation in the Lacandona forest and global change:

"Whenever there are talks, everyone's got their eye on us. Other countries are interested in us not destroying the forest. The danger is that it will become a desert and there won't be any rain or water or anyone;" and "As for the biosphere, as human beings, we are aware we shouldn't cut down the trees in the hills, because other countries need them."

Attitudes, Illness, and Overpopulation

Attitudes were considered one of the greatest dangers in all communities. For example, some respondents stated: "Humanity itself is wicked and perverse. People invent weapons, instead of helping the poor;" "Not obeying the government about not cutting down the trees. The government doesn't allow us to cut them down"; and "Drug addiction—that's something that terrifies me."[5] A mother in Lacandón said, without the slightest hesitation: "I know what worries me—it's the bars, where our children get drunk. That's what really gets me down."

When the answers are analysed by gender, we find that it is pre-dominantly women who refer to attitudes as one of the main dangers. This is understandable, as gender roles have placed women in these communities at the centre of moral issues and emotional relationships.

Illness was mentioned as a priority predominantly in communities with the worst pollution and the least access to medical services, including La Victoria and Nuevo Chihuahua in Marqués, Lacandón, La Unión, and among low-income groups in Palenque. Predictably, the high-income groups in Palenque did not refer to this factor.

Overpopulation was the fourth highest item of concern among high-income groups in Palenque and cattle ranchers. It was hardly recorded among the Marqués de Comillas communities and did not appear at all among the low-income groups in Palenque, El Lacandón, and La Unión. The expressions used in relation to this response and in many interviews were, for example, that "more and more people are always being born. There isn't enough room for them all"; and "We cause the danger ourselves; the more of us there are, the less work there is for everyone." In interviews this was reflected in statements such as the following by Genaro Zapata, a ranch owner:

> Here, nature is exuberant; it devours us. You can practically hear the grass growing. But we aren't able to exploit the forest; the problem is overpopulation. We need family planning programs, because there's no more forest left.

An ejido owner from Benemérito pointed out that

> the inhabitants of Mexico have already increased so much that what you produce isn't enough yourself or for the national markets. We've been having shortages in the region, because of the number of inhabitants. When there were only sixty of us ejido owners, we didn't have this sort of crisis.

Fieldwork showed, however, a generally passive attitude toward overpopulation, as exemplified by Santos Hernández, a milkman from the San Mateo ejido,

[5] Marijuana production has reached the Lacandona rainforest through drug traffickers, who have found the remote forest lots ideal for planting. It is said that it even has been planted in the Biosphere Reserve. Some farmers, although very few, are said to be involved, and certain *ejidos* in Marqués de Comillas are said to be centres of drug crops. This has led to forays and greater surveillance by the army. On its way from Palenque to the great markets of the north, marijuana has started to become easily available to the region's young people.

who commented: “God said that the Earth was going to get overpopulated and mankind was going to go hungry, and that’s exactly what’s happening.” This contrasts with less frequent statements reflecting concern about taking responsibility for such processes as overpopulation. For example, in the words of a civil servant from Palenque:

> The forest has a serious problem of overpopulation, and we have to do something about it fast. Families are very big, with eight, nine, or ten children, and it’s the church’s fault, with all that stuff about having all the children God wants you to have.

Interestingly, while women in the region do attend mass and consider themselves religious, their views on family planning seem to be diverging more and more. María González López, the wife of an ejido farmer from Nuevo Chihuahua, simply rejects this doctrine:

> [You have to have] the children God wants you to have? That’s not true. We can’t bring God into this because it’s a decision for each couple to make. My mother had eighteen children, but I had my tubes tied after the third. Times have changed; you can’t support them [children] the same way anymore. Men want to have children because they’re macho and to prove they’re men, but being a man means being a responsible father, although some of them don’t understand that, and then people say that a man who lets his wife have her tubes tied is an idiot because then his wife’s going to be unfaithful, and that’s not true either. Couples didn’t used to be like they are today; now they’re a real mess. Women don’t put up with things like they used to; the children are a disaster; and, when there’s a problem, it’s always the woman who has to sort things out.”

It is striking that more men than women cite overpopulation as a current threat—but with a difference. For women the issue is personal and domestic. For men it means that more migrants will be coming into the rain forest, as is continuing to happen, and pressure on the land, both in the settlements and for agriculture, will increase.

Answers related to divine punishment implied that God would soon be punishing humans because they are sinning so badly. The high percentage of this response in Nuevo Chihuahua came from a group of Seventh Day Adventists, who typically gave such answers as: “We always study the Holy Scriptures, and they tell us of the dangers there are, but the time of anguish and much enmity will come.” The response of a woman in La Unión was equally surprising: “If God decides that it’s all over, that would be even more dangerous than war.”

Other Answers

A few answers were given citing the main danger facing the world as the lack of technology, the possibility of natural disasters, the ban on tree felling, and the forest’s natural dangers. An older woman from La Victoria provides an interesting reflection:

> In my mind I can’t see anything [threatening]. In my time, when I was a little girl, there was a great shortage of corn, and now you don’t see that. Before people were hungry, now they are being productive. What I have noticed, though, is that there are more plagues—that’s what happening nowadays.

This reply is worth noting, since it was one of the few to assess current dangers in relation to a period in Mexico when there used to be famines. It would certainly be a pity, now that famine has been eradicated and people have land and seed, if the new threat today were extensive plagues, partly as a result of monoculture as a form of agricultural high- tech production, or the fall in food production because of soil depletion. Several people touched on these points, such as one who pointed out that "the land is no good for agriculture anymore. It's all burned up."

Only one person surveyed referred to the threat from other, more powerful nations: "As far as I can see, the greatest danger comes from the powerful countries, I mean the imperialists, as we poor go on being humiliated and exploited." The most fatalistic reply was: "Whatever happens, we can't do anything, so everything's dangerous."

6.3 Perceptions of Changes in the Natural Environment

The first difference we found is that, although only 6.7 percent of those surveyed cited deforestation as the main danger today, most of them perceived significant changes in the environment related to rain, heat, winds, floods, and the disappearance of animals, all of which are related to deforestation (see Table 6.4).

Answers to questions about environmental change, however, must be dealt with scepticism, since research has found that people often say that changes have occurred when this is not borne out by empirical data (Whyte/Anne 1985: 403). We have no reliable long-term data on rainfall, or other climatic phenomena, in the Palenque area and only personal or ethnographic statements that there are more droughts than before, more torrential rains, stronger winds, and more serious and frequent flooding in the forest. Therefore, we must treat these perceptions with reservation.

Table 6.4 Percentage distribution of those who believe that changes have occurred

Community	Rainfall	Heat	Winds	Floods	Animals
PR (1)	90	57	30	73	73
VC (2)	49	32	14	21	61
LU (3)	84	74	55	22	96
PL (4)	73	73	50	42	85
PH (5)	96	63	21	13	96
CR (6)	90	69	25	44	98
Total	78	59	33	37	82

Key 1 Pico de Oro and Reforma Agraria, *2* La Victoria and Nuevo Chihuahua, *3* Lacandón and La Unión, *4* Palenque low income, *5* Palenque high income, *6* Palenque cattle raisers

6.4 Which Environmental Changes are Most Often Perceived?

Of all questions asked in the survey, the one concerning changes in the natural environment had the greatest consensus among responses. Of those surveyed 82.2 % noted that there were fewer wild animals, 78.5 % cited changes in the level of rainfall, 59 % mentioned higher temperatures, 32.9 % noted changes in wind patterns, and 36.8 % named floods as a noticeable environmental change. A few interviewees referred to the entire range of natural changes in deistic terms such as one evangelist, who explained, "Throughout the centuries it was written that there would be changes wrought by God in the twentieth century."

Perceptions of the loss of biodiversity (including many kinds of birds, jaguars, and other wildlife) were predictably high in the already deforested areas around Palenque: more than 96 % among high- income groups from Palenque as well as the cattle ranchers and people in Lacandón. A cattle rancher from Palenque expressed this view eloquently:

> Now they've really destroyed everything! There's nothing left. I wish I'd had a camera to record it all. A man would ride along with his shotgun, and in a single day he would come back with a deer on either side [of the horse] and a boar on top of them. Another horse would be carrying a collared peccary and a spotted cavy. Many people made their living out of that.... But that's all gone now.

When asked how one could put a stop to this destruction, a blank look came over the man's face, and answered wearily, "Well, that's going to be very difficult."

In Palenque illegal animal trafficking was also mentioned: "I have a piece of the reserve in my ranch, but people come into hunt. In the beginning I had alligators, but they killed them and sold the skin, just like they did with the deer." Another explained: "People killed the animals because they [the government] didn't provide surveillance in time" Some regret this loss for economic reasons. In Palenque a waiter's wife said, "There aren't any more animals. There used to be plenty. We were really well-off because it was good, tasty meat. Now it's really expensive." A different, but interesting, remark was made by a woman from La Unión who said, "My hens and turkeys died from the heat."

In the rain forest communities, where wild animals are still to be seen, there is a widespread perception that the animals are disappearing, as shown in Table 4.4. Farmers in Pico de Oro commented some-what reticently, since they know they have been branded as the main predators. They said that the animals "had moved into the middle of the forest." Others, like the ejido farmer in Nuevo Chihuahua, were not so reserved: "The animals have disappeared. We know there's a ban, but, if we see one, bang!"—he gestures, making his hand into a pistol. Farmers in Reforma Agraria, who have undertaken measures for the animals' protection and are even experimenting with a program to protect the macaws, were cautiously optimistic: "The animals had begun to disappear, but maybe with the government ban there are going to be more."

Questioned about the rainfall, a high percentage—over 80 % in four of the six communities—had noticed changes. Unlike the question on animals, in which there was a clear difference between those in the forest communities and those living in Palenque, there was no distinction between the two in the question on rain. For example, inhabitants of the forest communities remarked: "Now the rain is unusual. It's badly distributed"; "Before it used to rain a lot, but the rivers didn't swell"; "When the mango trees blossom, the rains come and wash them away." In Palenque people commented that "before the bad weather used to last two weeks; now it don't last very long at all."

Some of the interviewees replied that it rained more now, while many others said that it rained less now. In Nuevo Chihuahua, for example, residents replied that "it used to rain a lot, but not anymore"; "It rains less, with eight months of drought"; "It's drier. In 1986 they burned down the forest, and in 1990 there was more rain." The same contrast was recorded in other communities. The explanation for this apparent contradiction was provided by a few interviewees in Reforma Agraria: "The seasons have changed. You never know when they're going to be anymore." In La Victoria some said: "The dates of the rains have changed. This time we had bad weather in December." In La Unión the response was similar: "Before we used to have more rain, but they've spread out a bit now. Before it used to rain in June, and now it doesn't rain until August, so we sow the seeds any time during the year." And, again, in the Lacandona forest people stated, "In June we used to get sudden storms, but not anymore."

These comments clarify that what has changed is the distribution of rainfall throughout the year. It rains less heavily than usual for several months but then rains torrentially during one or two months. Thus, there is a longer period of drought and a shorter period of intense rainfall, of the kind described in García Márquez's novel *One Hundred Years of Solitude* (1967). This double phenomenon is accurately reflected in the comment: "Either we run out of water, or it pours with rain and everything gets swept away."

As a result of these changes in rain distribution, the number of floods, especially those involving the forest's rivers, has risen. In December 1990, when we carried out our fieldwork, the Lacantún and Usumacinta Rivers had overflowed due to heavy rainfall in the Guatemalan highlands. Crops on the fertile lowlands of Pico de Oro and Reforma Agraria were swept away by the Lacantún River, which explains the high percentage of the following reply among members of these communities: "Before it used to rain a lot, but the rivers didn't rise." In the Palenque area inhabitants noted: "Now it rains an awful lot. The Chancalá River floods the motorway." Yet some of the interviewees pointed out that water storage has fallen to such an extent that, in the city of Palenque and the outlying areas, there are constant water shortages.

The response concerning higher temperatures was, not surprisingly, given most by those from the deforested area surrounding the city of Palenque (see Table 6.5). Over 72 percent of respondents in the five communities in this area reported a hotter climate, with comments such as, "Before the climate was balanced, but now

Table 6.5 Heat changes by region (percentage distribution)

Region	Yes	No	Noresponse
Forest	30.7	65.6	63.6
Palenque	61.3	34.5	35.4
Total	100.0	100.0	100.0

it's much hotter" and "The sun's rays burn more, and eight or ten years ago didn't" (Table 6.5).

Finally, no clear pattern emerges among responses regarding changes in wind patterns. This could be interpreted to suggest that the two rural communities in Palenque (i.e., the two in which deforestation has been greatest) are the most severely affected by the winds, while changes in the winds are only dimly perceived by the other groups. People mention most frequently that the mango crop is not as abundant as it used to be because today's rough winds prevent trees from being pollinated. In these responses there are practically no differences in perception between men and women, mestizos and Indians, or among religious groups. In the face of clear, empirical phenomena, generalized perceptions are beginning to emerge, in which cognitive or cultural differences regarding the environment are no longer relevant. One could say that, as a consequence of ecological phenomena becoming increasingly visible and more generalized in scope, the semantic field of perception of these phenomena, which is highly divided at present, will become unified. Those in rural communities perceive these phenomena most directly, whereas in the urban community the perception is modified by television and radio.

As a last point, it should be pointed out that very few of the interviewees established a link between these changes and deforestation. Only one cattle rancher expressed the connection explicitly: "Deforestation has changed everything." As mentioned, Protestants tend to associate these changes with God's will. The majority of the interviewees, and the area's population in general, did not associate the disappearance of the forest with climatic changes or the loss of bio-diversity. They associate the loss of animals with hunting for food by the peasants and with smuggling, but not with the disappearance of the forest. Conversely, the destruction of the forest is associated with the loss of firewood, timber for building, shade, and, finally, the countryside.

References

Arizpe, Lourdes; Paz, Fernanda; Velazquez, Margarita, 1996: *Culture and Global Change: Social Perceptions of Deforestation in the Lacandona Rain Forest in Mexico* (Ann Arbor: The University of Michigan Press).

Derrida, Jacques, 1978: *Writing and Difference* (London: Routledge and Jegan Paul).

García Márquez, Gabriel, 1967: *Cien Años de Soledad* (Buenos Aires: Sudamericana).

Lévi- Strauss, Claude, 1973: *Antropologie Structurale* (Paris: Plon).
RAE (Royal Academy of Spanish), 1984: *Diccionario de la Real Academia de la Lengua Española* (Madrid: Espasa-Calate).
Whyte, Anne V.T., 1985: "Perception", in: Kates, R.W.; Ausbuel, J.H.; Berbertian, M. (Eds.): *Climate Impact Assessment* (Toronto: SCOPE, John Wiley and Sons): 403–436.

Lourdes Arizpe's Ph.D. thesis for LSE on Migration, Ethnicism and Economic Change that was published in Spanish in 1978

Chapter 7
Relay Migration and the Survival of the Peasant Household

7.1 Introduction

The massive labour migrations of modern history have consistently been associated with the geographical mobility of capital.[1] The unequal distribution of capital investment around the world has generated pervasive unequal development between rural and urban areas, among regions within a country and among countries. Within this context, rural–urban migration in developing countries can be considered as the geographical counterpart of the generalized exchange of labour and resources between the peasant and the industrial sectors, an unequal exchange that has greatly benefited the urban centres where industry is located. In analysing this movement, much attention has been given to economic and social pressures driving peasants from their land, or attracting them to the cities, but few efforts have been made to understand the reactions of peasants to such pressures, that is, their strategies to overcome encroaching poverty and deprivation.

This chapter analyses data from Mexican peasant communities and attempts to explain how peasant households use migration as a strategy for survival and reproduction. This strategy is important in clarifying the role played by peasant communities (and within them by rapid population growth) in capitalist development.

Migration studies in Latin America have identified the major characteristics of migratory cohorts (Elizaga 1970; Colegio de México 1970) as well as the structural processes that link industrial growth and urbanization (Singer 1975). In some countries, the patterns of geographic mobility, the fluctuations of migration during different historical periods, and the adaptation of migrants to urban society have also been analysed (Balán/Browning/Jelin 1973; Muñoz/de Oliveira/Stern 1977; CLACSO 1973, 1975).

---

[1] This text was first published in: Jorge Balan (Ed.): *Why People Move* (Paris: UNESCO 1982): 187–210; it was republished in Helen Safa (Ed.): *Toward a Political Economy of Urbanization in Third World Countries* (New Delhi: Oxford University Press, 1982). This text is used with permission of UNESCO and the permission was granted on 4 July 2013 by Isabelle Nonain-Semelin on behalf of UNESCO.

L. Arizpe, *Lourdes Arizpe*, SpringerBriefs on Pioneers in Science and Practice 10,
DOI: 10.1007/978-3-319-01896-6_7,

It has been far more difficult to explain the complex factors associated with migration in peasant communities. Part of the problem has arisen from the failure to distinguish between migration and rural unemployment at a theoretical level. Because attention has been focused primarily on structural processes, it has been taken for granted that the same mechanisms that explain the rise of a relative surplus population in rural areas also explain the rural exodus. It is now clear that the problem of selectivity does not allow for such an easy generalization. In this chapter, the view is taken that structural processes provide the necessary conditions for migration—as Clyde Mitchell has stated—but that the patterns of age, sex, recurrence, and occupational specialization can only be understood with reference to the labour and social conditions of the peasant household. In other words, the needs and resources of the households are the immediate framework within which personal decisions to migrate are taken.

The relationship between unemployment and out-migration varies according to social class in rural communities. Agricultural labourers feel the impact of fluctuations in the labour market more directly and are more prone to migrate as a result of unemployment. At the other end of the scale wealthy peasants can fall back on their resources so that their sons and daughters migrate primarily in search of educational and social mobility. The causes of migration are far more complex for the largest rural group, that of the small landholding peasants. They migrate as a response to rural unemployment as well as to contribute to the household income or to attain social mobility.

The proposition that each social class tends to generate specific forms of out-migration in response to the economic and social pressures that affect it summarizes the results of previous research carried out by the author (Arizpe 1978). That analysis showed that the most complex and interesting questions referred to the survival and reproduction of the small landholding peasants. Faced with economic pressures that tend to expel them from traditional agriculture, how have they managed to persist? In contrast to European history, where the industrial proletariat was the most prolific class, why is the impoverished peasant in Latin America as well as in other developing regions the group with the highest rate of population growth? There is a high incidence among such peasants of permanent as well as recurrent and temporary out-migration, but what is the relationship between these different forms of migration?

In an attempt to answer these questions, it has been argued that the pre-capitalist peasant economy continues to exist because the capitalist sector needs to extract labour and resources from it constantly in order to reproduce itself (Meillassoux 1977; Castells 1975). To show how these resources are being transferred to the capitalist sector attention has been focused on the peasant unit of production and reproduction, that is, the household.[2]

[2] Interest on this subject revived after the work by Chayanov (1966), Meillassoux (1977) contributed the concept of domestic economy, while Tilly and Scott (1978) provide important historical data and a new perspective.

The significance of the household as a unit of analysis has also been demonstrated by recent research on rural exodus and on the role of female labour in Europe in the eighteenth and nineteenth centuries. Studies on these subjects have shown that the sole male heir pattern of inheritance in Austria, western Czechoslavakia and England favoured a high degree of geographical mobility of labour which in turn allowed a highly centralized pattern of industrialization to develop (Habakkuk 1960). According to Frédéric LePlay (1871), the type of family that resulted from, this inheritance pattern, the *famille souche* or stem family, was well suited to economic development, because it kept family capital intact and allowed the heir to re-invest. Also his brothers and sisters could return if they were unable to survive in the city, thus saving public expenditure in the city. What LePlay overlooked is the relationship between family composition and social class. The strategy he refers to, of sending 'branches', brothers and sisters, to other sectors of the economy, can only be used in families with considerable wealth.

Several recent studies have shown that wealthy rural families tend to organize themselves into extended families (Berkner 1972; Young 1978; Deere 1978). However, it is useful to distinguish between stem family and extended family. The latter is a household made up of more than two linear generations and/or first-degree relatives. In contrast, the stem family is a type of extended family that implements the strategy mentioned by LePlay.

Berkner (1972) argues that the increasing prevalence of nuclear families during the period of industrial growth in Europe in the nineteenth century was not due to a change in the cultural norm of family composition but to the relative increase of the salaried classes where this norm is predominant.

What is the relationship between family composition among small land-holding peasants, and their survival in a market economy? The answer to this question is related to the availability of resources to which such house-holds have access. It is a well-known fact that peasant societies the world over are losing resources constantly through price mechanisms, taxes, interests on loans and other channels. How is this loss compensated for so that they are able to reproduce their family agricultural enterprise? It seems that in France and other European countries this compensation is provided through debts to the state. Where does it come from in Latin America?

This chapter suggests a partial answer to the above: part of this compensation comes from the strategy of relay migration. By sending the father and later the offspring in progressive order to work in wage employment, the peasant household obtains resources which enable it to continue in agriculture and which ensure its social reproduction.

7.2 Toxi and Dotejiare

At the beginning of this century, the Mazahua Indian region in which the villages of Santiago Toxi and San Francisco Dotejiare are located was dominated by haciendas (Arizpe 1978). These large landholdings produced maize, beans, wheat, and green peas, and specialized in cattle breeding and in the export of *zacatón,* a fibrous root used in the manufacture of brooms and scrubbing brushes. Peasant communities were interspersed in this region of lxtlahuaca and Atlacomulco. In contrast to Morelos, where the expansion of the sugar industry caused the constant usurpation of peasant lands, there was no direct confrontation between peasant communities and the haciendas, in this region, since hacienda production was limited by national and international market conditions for its products. Thus, in the revolutionary movement of 1910, instead of uniting against the hacienda owners, the peons and peasants of this region were divided among themselves according to their loyalties to local caudillos, or regional leaders.

As a result, once the armed struggle was over, the haciendas reorganized and continued to produce as if nothing had happened. It was not until the end of the 1920s that *agrarista* groups appeared demanding land reform and, encouraged by governmental redistribution of other haciendas, began to invade the large landholdings. A few years later, agrarian reform was carried out in the whole region. With the granting of *ejido* land to the peons, land owned by the state but held in usufruct by the peasants, a peasant economy was consolidated in the region. This was an economy based on domestic production but fully integrated into the national market through the sale of agricultural goods and handicrafts, and into the international market through the sale of *zacatón* root. From that time on, this peasant economy has been deteriorating. In the case of the majority of families with *minifundio,* the result has been acute poverty and massive emigration to Mexico City. A few families have been able to expand their properties and gain economic and political hegemony over the region, serving as intermediaries between the local producers and the national market.

In Santiago Toxi, located 250 km north-west of Mexico City, the *ejido* was formed in 1928, granting an average of 2.5 ha of land to each of the 718 male heads of households. This was considerably less than the four hectares considered as the minimum necessary holding for the subsistence of a family at that time. The small size of these landholdings was due to the relatively high population density (sixty-one inhabitants per square kilometre in 1930) since many haciendas and peasant communities had concentrated in the fertile and centrally located Ixtlahuaca valley. Although the land allotments did not always follow the patterns of residence or of ethnic groups, the peasants who were freed from the control of the haciendas tended to group together on the basis of ethnic ties. All the approximately 300 families, who settled in Toxi, spoke the Mazahua Indian language and had previously lived and worked on the hacienda lands.

In San Francisco Dotejiare, another community 50 km northwest of Toxi, 451 families received *ejido* lots that averaged 6.2 ha. Population density was low and the village had not been encroached on by the nearest hacienda.

Throughout Mexico at this time it was not clear how the newly created *ejido* system would organize production. In the cases of Dotejiare and Toxi, formal collective land grants were made, without subdividing the property into individual holdings. The specific distribution of piers of land to each member of the *ejido,* or *ejidatario,* was left to the arbitration of the local *ejido* commissary. But once the land had been distributed, the state failed to provide access to the technical, administrative and credit channels that would have supported either collective or co-operative production. For their part the new *ejidatarios* were not interested in developing such channels. In Toxi, their request for the immediate distribution of titles to individual lots was granted.

The situation in Dotejiare was made more difficult by the presence of *zacatón* plantations on *ejido* lands, which led to internal conflicts lasting until 1952. One of the factions in this struggle demanded the immediate subdivision of the land to allow for *zacatón* production on an individual basis. This system, which was adopted by a nearby *ejido,* had disastrous results because the *ejidatarios* over-cultivated the root, ruining the plants and causing uncontrollable erosion. The second faction wanted to maintain the communal cultivation of the plant. The final triumph of the second group has made Dotejiare the only village in the area that still makes considerable profit from extensive *zacatón* cultivation.

The fact that there were no specific boundaries for individual land plots in Dotejiare also had important consequences for family labour. Capital was scarce and produced no variation in the size of cultivated area, while the availability of family labour was essential. Families with several sons were able to expand the cultivated area (according to the law, daughters could not inherit *ejido* property). In addition, family labour could be occupied in profitable domestic industries. Handicrafts, cottage industries and trade allowed peasant households to diversify their incomes.

These factors, along with others such as the improvement in the nourishment of mothers and children, contributed to a marked population increase in both villages. Between 1930 and 1940, the average rate of annual population increase was 2.8 per cent in Toxi and 2.9 per cent in Dotejiare. Even so, mortality, particularly infant mortality, was very high.

Data from witnesses shows how the division of labour by sex and age within the households changed in response to the new economic opportunities. The father and elder sons were responsible for the agricultural tasks, with the women's assistance during planting, weeding and harvesting. The fact that earlier forms of labour exchange among groups of relatives disappeared indicates that the domestic groups tended to be self-sufficient in these activities.

This self-sufficiency was made possible by the availability of local and regional sources of cash income which enabled the families to rent a team of oxen and a plough. The father and sons obtained this income by working part-time in the El Oro mines and on neighbouring ranches. Some also traded throughout the region. Those from Dotejiare were commonly known as the 'fruit men' because they sold fruit that they brought from Zitácuaro, three days away on foot. In Toxi, there were 'chicken men' who carried fowl to the central market of Mexico City. The inhabitants of both villages occasionally carried a variety of merchandise by mule to be traded in other regions. In addition, in Dotejiare the men cultivated the *zacatón* root and prepared the alcoholic beverage known as pulque from the maguey cactus to earn extra money.

They also wove sarapes, or woollen blankets, for sale. In Toxi, the inhabitants wove grass and rush mats known as *petates* for sale.

As part of their domestic labour the mother and elder daughters prepared and cooked food, looked after the children and raised domestic fowl, in addition to the usual household tasks of carrying water, washing clothes, mending, etc. They gathered edible plants, honey, and other wild foods. It was the women who took charge of the small local trade, carrying farm produce from the fields and animals from their household pens to markets in the nearby villages. As for crafts, the women of Dotejiare made most of the family's clothing and were specialists in embroidery, while the women of Toxi wove woollen articles and made some clothing. The younger daughters and sons cared for the sheep. The sons also gathered wood and helped their fathers in agricultural or craftwork. The younger daughters looked after younger brothers and sisters and helped their mothers with household chores. Thus, the households were able to make maximum use of their available labour. This is reflected in the fact that until 1940 there was no out-migration from the villages, except for the seasonal migration of the father and older sons to wage labour within the same region.

In the 1940s the second generation of *ejidatarios* began to inherit the land. By that time, the *ejido* plots in Toxi had been distributed ad hoc to *ejidatarios* and there was no possibility of enlarging them other than by purchasing private lands in the centre of the village. This could still be done in Dotejiare, using the profits from the sale of *zacatón* root to acquire new lands. But in Toxi, the young men began to inherit plots so small that they could not support a family.

They began to migrate on a temporary basis, some to other agricultural zones, and others to Mexico City to work in the central market. Some of the women also began to go to the city to seek employment as paid domestic workers. A few men and women took up permanent residence in the city, but the majority returned to Toxi or moved to nearby villages. Vital statistics indicate that between 1940 and 1950, 25 per cent of the female population (414 women) and 21 per cent of the male population (351 men), left Toxi.

There was much less out-migration from Dotejiare during the 1940s; only 279 women and 208 men left (10 and 16 per cent, respectively, of the population). Almost all of the male migrants went to work in the vegetable gardens of Xochimilco, a suburban agricultural zone of Mexico City. A few others sought employment extracting *zacatón* root in the states of Puebla, Michoacán and Jalisco. The women who left Dotejiare were not looking for jobs, according to local sources; more probably they married and moved out to other villages.

Between 1950 and 1960 the natural population increase was slightly higher than the national average. Because of specific circumstances this demographic pressure was not felt as acutely in Dotejiare as in Toxi. In 1950, close to twenty-five Dotejiare families migrated to Mexico City as a result of disputes over the control of the *ejido*. Consequently, a survey made in 1956 showed that each *ejidatario* still had an average of 2.5 ha of land. In contrast, the redistribution of land by inheritance in Toxi had reduced the *ejido* plots to an average of only one hectare by 1956. Consequently, out-migration, this time almost exclusively to Mexico City, became an integral part of community life. Decisive economic changes had occurred, altering the balance of income and expenditures of the peasant households.

With the fragmentation of the *ejido* allotments and intensive cultivation, the land rapidly lost its fertility. In 1956, one hectare of non-irrigated land barely produced 600 tons of corn. Since the 1960s, the lands have given no yields without the application of chemical fertilizers (animal fertilizer is no longer available). Nor is this the only cash outlay that families must make on their crops. In addition most of them must rent either a team to pull the plough, or a tractor, and must pay wages for temporary workers. Because of the rising costs of agricultural day labour, families benefit from having several sons and daughters who migrate only seasonally, and can therefore help with weeding and harvesting when they are needed. It is estimated that the total cost of corn production doubled between 1950 and 1970. At the present time, 80 per cent of these production costs require cash payments (Díaz 1974).

Another factor that has contributed to the scarcity of land is the drastic soil erosion that has resulted from rapid deforestation. Wood is used both as fuel and construction material for the houses of Dotejiare. No efforts have yet been made to reverse the effects of uncontrolled cutting of the trees, bushes and *zacatón* grass that once covered the region. In Toxi, between 1927 and 1950, 21 ha of *ejido* lands were lost due to erosion. By 1960, another 72 ha had been destroyed.

Possibilities for local employment have decreased. The foreign companies that owned the gold mines of El Oro began to close them down one by one, until the last one stopped working in 1954. Mechanization of the largest ranches has reduced the demand for agricultural labourers. In the 1940s men from Toxi found employment in the construction of gravel roads and paved highways throughout the region. Since their salaries did not include payment for food or lodging, food was brought to them from their home villages and they slept in the ditches they had dug during the day. In the 1950s, even this means of gaining a cash income disappeared.

Traditional occupations also began to disappear. The local folk-doctor and midwife were replaced by doctors from the city. People no longer hired the village fiddler or guitar player, but brought musical bands from the city. The firecracker-maker, the altar decorator, the teacher of traditional dances, the medicine woman, the roofer, and the prayer man, have all lost their jobs.

Small local business and itinerant trading have also declined. They have been taken over by the large commercial enterprises in the regional cities and incorporated into the monopolistic activities of merchants from the central market, La Merced, in Mexico City. Handicrafts and cottage industries such as the brewing of local liquor have declined almost to nothing because of competition with manufactured products from urban industries. Not until the 1970s did the state government begin to take an interest in reviving traditional craft industries, but they have still not regained the importance they had as a source of income before the 1950s.

Two important activities still exist in Dotejiare. *Zacatón* root continues to be exported and to provide families with a consistent cash income; in spite of the fact that exports are down to a quarter of what they were at the beginning of the century.

Also its price has declined drastically as a result of international competition from synthetic fibres. Pulque, the alcoholic beverage extracted from the maguey cactus, is still produced and sold, though in much smaller quantities than before, as soft drinks and bottled beer now dominate the market. The price of pulque is, proportionately, 20 per cent of its price of thirty years ago.

The changes so far described were destroying the means by which peasant families had once been able to obtain a cash income. At the same time, the cash needs of such families for non-agricultural purposes increased. Electricity, the ever more frequent bus journeys, the extras required at school, although education is free, and, in Toxi, drinking and irrigation water are all new services that must be paid for. In addition, consumer habits which are constantly advocated by mass advertising have notably increased the acquisition of goods such as radios, watches, manufactured clothing, furniture and record-players.

According to the national census of 1970, both Toxi and Dotejiare are considered peasant communities, since 67 and 88 per cent respectively of the heads of families work in agriculture.[3]

As far as the father's occupation is concerned the results of the household survey in the villages resemble those of the census, as shown in Table 7.1. However, the survey also shows that the majority are engaged not only in agricultural work, but combine it with some other activity.

According to the preceding figures, 82 per cent of the cases in Toxi and 95 per cent of those in Dotejiare are involved in agriculture, which would lead to the conclusion that these communities are primarily peasant economies. There is, however, a different conclusion if the activities of the entire household labour force are analysed, that is, taking into account all working adults, both women and men.

Table 7.2 shows the distortion that results from classifying the household as a peasant one on the exclusive basis of the father's occupation. It becomes clear that wage labour is the predominant occupation in Toxi. In Dotejiare, only 12 per cent of the population is dependent on a wage. Toxi, then, in spite of appearances, is no longer a peasant community.

By examining the distribution of labour inside the households, it is possible to evaluate both their capacity to absorb labour and their need for cash incomes from wage labour. In Toxi, 53 per cent of the households have one or more members who work in agriculture; in Dotejiare the figure is 96 per cent. Of these, 54 per cent, in Dotejiare, have a single agricultural worker, 32 per cent have two and the remaining families have three. The latter includes two families with medium-sized land holdings and seven that are large landowners. In these cases, the sons do not actually work as farm labourers, but oversee the hired help or tend the family store. This is in marked contrast to Toxi, where 92 per cent of the households having agricultural workers have only one such labourer and 8 per cent have three. These last cases are the result of special circumstances. In two of them, for personal reasons, the father has refused to allow his sons and daughters to leave the village

[3] Data obtained by special request from the Department of Statistics.

Table 7.1 Distribution of main occupations of fathers

Occupation	Toxi No. of house-holds	%	Dotejiare No. of house-holds	%
Agriculture only	16	25	6	10
Agriculture and extractive activities (pulque or *zacatón*)	17	27	52	85
Subsistence agriculture and migratory				
Wage labour	19	30	–	–
Local wage labour only[a]	7	11	1	2
Migratory wage labour only[b]	4	6	1	2
Total	63	100	60	100

Source Survey (1976)
Cases not included: Toxi, II; Dotejiare, I0
[a] Includes farm labour, construction work, administrative and service activities
[b] Includes informal employment, construction work, administrative activities and self-employed activities in the city

Table 7.2 Main occupation of all household workers

	Unpaid							
	Agriculture[a]		Domestic Labour		Wage		Total	
	No. of house-holds	%	No. of house-holds	%	No. of house-holds	%	No. of house-holds	%
Toxi	45	21	69	32	100	47	214	100
Dotejiare	107	53	70	35	25	12	202	100

Source Survey (1976)
[a] Includes extractive activities and two cases of small businesses in agricultural products

in search of work and has forced them to work as labourers in nearby fields. Another case is that of three brothers who, after the death of their father, have continued to live in the family home with their wives and children, cultivating their land and occasionally hiring themselves out to the neighbouring landholders. The figures suggest that in the majority of households in both villages there is only enough work in the family fields to occupy a single full-time worker. In Dotejiare the exploitation of *zacatón* and pulque allows a second full-time worker in one third of the households.

However, the crops require intensive labour during seeding, weeding and harvesting. The additional labour for these agricultural tasks is provided by family members who migrate temporarily, and by unpaid female domestic workers.

In 90 per cent of the cases in both villages, households include unpaid female domestic workers. The remaining 10 per cent are special cases of widowers. How many female workers are there in each household? In 80 per cent of the cases in Dotejiare, and 94 per cent in Toxi, there is only one such worker. This indicates that a large number of elder daughters who would otherwise appear in the survey, have left home. Indeed, among the first and second children remaining in the domestic units there is a much higher proportion of males than females.

The fact that in Dotejiare a greater number of cases have more than one domestic worker cannot be explained by the consideration that larger families require more help. In only a few cases of wealthy families is the daughters' help required in order to run the family store. On the contrary, households in Toxi are larger, which means that the single domestic worker in Toxi has a heavier burden of domestic activities, made even heavier by the fact that the younger daughters now attend school.

If the number of women who leave Toxi to work as paid domestic workers in Mexico City is added to the number of women who remain working in the homes in the village, the figure is very close to the total number of domestic workers in Dotejiare. This indicates that Dotejiare families do not have the same urgent need for the additional cash income that forces Toxi families to send their daughters to work in the city. Indeed, residents of Dotejiare consistently considered it humiliating for a Mazahua woman to work as a servant. For this reason, the few women who do migrate to the city usually become fruit vendors.

Wage labourers were present in 89 per cent of Toxi households and in 28 per cent in Dotejiare; of those in Toxi 59 per cent have one worker, 30 per cent have two, and the remaining 11 per cent have three. In Dotejiare 81 per cent have one wage labourer and the remaining 19 per cent have two. This contrast is very significant; it shows the need for multiple cash incomes in Toxi.

It is even more revealing if the place of work and the type of wage employment are analysed.

Table 7.3 confirms that possibilities for wage employment in the area are disappearing and have not been replaced by new ones. New jobs have been created in the area through the slow expansion of administrative, financial and business services in the small regional cities. However, most of these jobs are taken by those who live in these cities, and only a few people from the villages have been able to fill such jobs. In fact, even young people from these regional cities have not been able to find employment and have also migrated to Mexico City.

A few family-owned businesses such as bakeries and small stores only provide jobs for members of their own households. However, this is not always an easy matter. On two occasions, small factories had to shut down because they were unable to retain a stable core of workers: the latter preferred to migrate to higher wages in Mexico City.

The only factory that has been successful in creating a steady work force is an electrical-appliance factory that employs close to fifty people from Toxi. It has survived because of large capital investments that enabled it to sustain the losses of the first few years, and also because of an intensive campaign to promote cultural values appropriate to factory work and consumerism. It was also able cut down costs by replacing its original male work-force with women. According to the manager, women are 'not as restless' and 'more dexterous in detailed hand work', than men. It may also be added that women are paid consistently lower salaries than men, are assumed to be more submissive to authority and less likely to

Table 7.3 Percentage of local and migratory wage labour

	Local[a]		Migratory[b]		Total	
	No. of households	%	No. of households	%	No. of households	%
Toxi	14	14	86	86	100	47
Dotejiare	5	20	20	80	25	12

Source Survey (1976)

[a] Includes farm labour, factory jobs and administrative and service work in the area

[b] Includes informal occupations, factory and administrative work and self-employed activities in the city

unionize. They are dismissed when they marry and this enables employers to avoid the costs of maternity payments (required by Mexican law), salary increases according to seniority, and old-age pensions. Nearly 80 per cent of the 2,300 workers in this factory are women.

The possibility of employment in this factory has kept many young people from leaving Toxi in search of work. Indeed, 85 per cent of the local wage labourers represented in Table 7.3 work in this factory. If this work were not available, predictably 94 per cent of wage labourers would be migratory.

In Toxi 40 per cent of the total labour force is involved in temporary migration. The percentage is even more dramatic when those workers, whom the Mexican census considers 'economically active', are taken into account, that is, if all unpaid female domestic workers are included. The result would be that 60 per cent of the 'economically active' work-force of Toxi is involved in some form of temporary migration. Therefore the village could be more accurately characterized as a proletarian suburb of Mexico City.

By contrast, in Dotejiare, only 10 per cent of the workers (15 per cent of the 'economically active') migrate temporarily to the city. In addition, the composition of the groups of migrants of each village is quite different according to sex and to the type of activity in which they engage in the city. Of the migrants from Toxi, 84 per cent are men who arc employed as unskilled construction workers, loaders and stock workers in the market-places or as store and warehouse employees. The remaining 16 per cent are women, all of whom are employed in paid domestic service and send money home to their families. In contrast, 60 per cent of the migrants from Dotejiare are women who are primarily fruit-sellers and only a few of them send money to their families in the village.

Why is temporary migration so high in Toxi? If men are aware that even if they do receive land, the plot will be far too small to make a living from, and if women know that if they do not marry they will not be able to remain in the village, why do they not migrate permanently to the city? One possible answer is that they have not been able to find steady jobs in the city. The majority are, in fact, underemployed. But data showed that the older brothers of those who were migrating at the time of our survey were well established in the city in permanent jobs. Also, among those who migrate only temporarily, many have steady well-paid jobs as factory workers or office employees.

One suggestion is that these migrants have not severed their links with their parents' household because their migration is functional to the household. In other words, they are fulfilling the role they have been assigned within the division of labour in the peasant household. In order to understand how these roles are assigned this chapter goes on to analyse the relation between migratory wage labour and the composition of the peasant families.

7.3 Division of Labour and Family Composition

The average number of members per household is slightly higher in Toxi (6.8, with a standard deviation of 2.4) than in Dotejiare (6.2, with a standard deviation of 2.2). Table 7.4 shows the distribution of families according to size in the two villages.

The larger households in Toxi are not due simply to extended families. On the contrary, only a third (21 cases) of the domestic units in Dotejiare and only 19 per cent (14 cases) of those in Toxi could be considered extended families. In almost all cases of extended families in both villages, agriculture is the main occupation of the father. In households where the father is a wage labourer no cases of extended families were found. In addition, the majority of the extended families belong to the wealthiest group in both villages. The number of cases of extended families in our sample (10 in Toxi and 16 in Dotejiare) is too small to allow for generalization, but it supports the hypothesis of the relation between composition of extended families, the type of economic activity they practice and the size of their landholdings. Half of the extended families in this survey include daughters-in-law and a fifth of the sample has sons-in-law. Of the remaining cases, two are composed of wives and offspring of several married brothers, two others include grandmothers, two include nieces and nephews and one case includes an adopted child.

The presence of the wives of young married sons in the household is a normal stage in the domestic cycle (Goody 1968; Arizpe 1972). Figure 7.1 shows that sons and daughters-in law most frequently appear in the stage of the domestic cycle in which the mother is between 39 and 50 years old. The decline in number of members in the household when the mother is between 31 and 38 years old is

Table 7.4 Distribution of domestic groups according to size

	Number of members							
	1–5		6–9		10–13		Total	
	No. of households	%	No. of households	%	No. of households	%	No. of households	%
Toxi	18	25	44	61	10	13	72	100
Dotejiare	29	42	32	47	7	10	68	100

Cases not included: Toxi, 2; Dotejiare, 2

explained by the fact that many young couples spend the first few years of married life in the city. They return to the parental home when they begin to raise their children. Figure 7.1 also shows that in Dotejiare there are more extended families throughout the entire cycle, particularly during the last stages. In Toxi the decline during the last stage, when the mother is over 50, indicates that only the youngest son remains in the house.

The family property, in fact, is traditionally inherited by the youngest son. Women are excluded from the inheritance, as it is expected that they will marry. Although some brothers continue to support unmarried sisters, the fact that they do not inherit directly reinforces the economic and social hardships of unmarried, widowed, or divorced women. It is important to note that the norm of sole heir seems to apply only when the elder brothers are able to establish themselves successfully elsewhere. If there is enough land, or if they are unsuccessful, the father usually divides the land among all the sons. Between 1930 and 1950, landholdings were distributed in this way. While this pattern was followed in Dotejiare until the 1970s, from 1950 onwards families in Toxi began to encourage their older sons to seek occupations outside agriculture. The expansion of the job market in Mexico City between 1950 and 1965 made it possible for these sons to enter the urban economy successfully.

Figure 7.2, on the other hand, shows that despite the scarcity of land and the declining income of the families in Toxi there are more children in the household

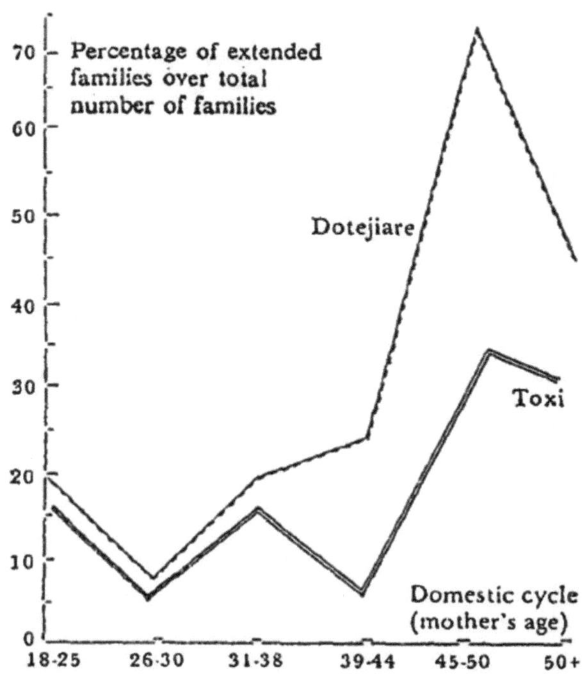

Fig. 7.1 Percentage of extended families according to domestic cycle

throughout the entire domestic cycle than in Dotejiare. Although, as indicated in the figure, women of Dotejiare tend to begin having children at a younger age, the women of Toxi have a larger number of offspring.This discrepancy cannot be explained by infant mortality since nutritional standards are the same and similar medical and health services are available in both communities.

While in Toxi the average number of resident children per family is 4.51 in Dotejiare it is only 3.6. Toxi, in fact, has a higher birth rate than Dotejiare. However, such data must be verified with appropriate demographic techniques. In this chapter, the analysis centres on the fact that families in Toxi seem to retain their children in the parental household for a longer period than in Toxi. This may be because of the specific role they play in the household as a means of counteracting the declining level of income. This role varies in each stage of the domestic cycle.

Table 7.5 shows the proportion of workers to dependents according to the stages of the domestic cycle in the households.

The critical moments in the ratio clearly correspond to the stages at which the children have not yet begun to work full time. In Dotejiare this stage is a little earlier than in Toxi, given that people marry at a younger age, but the ratio in Dotejiare is never as unbalanced as it appears in Toxi in the second and third stages of the domestic cycle. At these two stages, there are more than two dependents for each worker. What types or work do these workers carry out during the various stages?

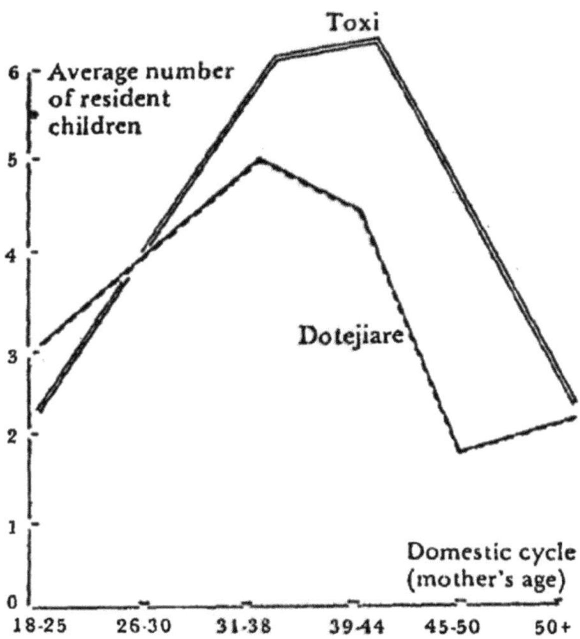

Fig. 7.2 Average number of resident children according to domestic cycle

Table 7.5 Ratio of workers[a] to dependents in the domestic cycle

Domestic cycle (mother's age)	Toxi	Dotejiare
18–25	1.16	1.55
26–30	2.06	1.77
31–38	3.29	1.77
39–44	1.48	0.95
45–50	1.05	0.96
50+	0.78	1.10

Cases not included in each village: 4 or 2.8 per cent
[a] Includes self-employed agricultural workers, domestic workers and wage labourers

Table 7.6 shows that migratory wage labour is continuous in Toxi households throughout the entire cycle. This contrasts with Dotejiare where the activity is concentrated in the fourth and fifth stages. A comparison with the previous table confirms that labour migration in Dotejiare is not an economic necessity, since it occurs at the stages where the ratio of workers to dependents is most balanced.

Table 7.6 also makes it clear that young couples in Toxi depend almost exclusively on migration for their livelihood. Indeed labour migration in Toxi is more important than agriculture at all stages, while the opposite is true of Dotejiare.

Table 7.6 Distribution of types of work by household workers[a] during the domestic cycle (percentage)

Domestic cycle (mother's age)	Agriculture	Unpaid domestic labour	Local wage labour	Migratory wage labour
18–25				
Toxi	0	42	25	33
Dotejiare	50	45	5	0
26–30				
Toxi	7	47	3	43
Dotejiare	51	38	4	7
31–38				
Toxi	18	39	8	35
Dotejiare	61	34	0	5
39–44				
Toxi	27	30	9	34
Dotejiare	61	26	0	13
45–50				
Toxi	22	30	7	41
Dotejiare	43	32	6	18
50 or over				
Toxi	35	19	10	35
Dotejiare	54	32	4	10

Cases not included in each village: 4 or 2.8 per cent
[a] Number of workers per specific activity in the household, over the total number of workers in the household. This ratio is averaged for ail households in each stage

Unpaid domestic work also shows the highest rates during the first stages, before the mother has reached the age of 38. It is precisely during these first stages that women intermittently engage in wage labour outside the home but near the village, as servants, agricultural labourers, washer-women, or small traders.

In Toxi the majority of the migrants are men, while in Dotejiare almost all are women. But who are these migrants according to their position within the household? This is shown in Table 7.7.

There is no correlation between position within the household and migration in Dotejiare, whereas in Toxi a clear pattern emerges: first the father and then the sons and daughters progressively, as they grow up, migrate each in turn. This pattern has been called relay migration.

This type of migration is a response to specific needs of households during the domestic cycle. For Toxi families, the critical stages in terms of the balance between workers and dependents occurs around the time the mother is 25 until she has reached the age of 45. During these twenty years at least one cash income, and sometimes several, is ensured by the wage labour of some member of the family. To achieve this, at least three or four children are necessary if the father is to be substituted for through migration.

Migrant daughters generally contribute to the family income for a period of less than five years, since they usually migrate around the age of 14 and marry between 18 and 20. The first few years after a woman is married her income generally goes

Table 7.7 Percentage of members of the household who migrate according to the domestic cycle

Domestic cycle (mother's age)	Father	First child[a]	Second child	Third child	Others[b]
18–25					
Toxi	84	–	–	–	16
Dotejiare	–	–	–	–	–
26–30					
Toxi	100	–	–	–	–
Dotejiare	–	80	–	–	20
31–38					
Toxi	66	29	5	–	–
Dotejiare	33	33	33	–	–
39–44					
Toxi	9	43	43	5	–
Dotejiare	-	84	16	-	–
45–50					
Toxi	8	55	21	11	5
Dotejiare	–	70	–	10	20
50 and over					
Toxi	4	61	23	12	–
Dotejiare	17	50	17	16	–

[a] Refers to the oldest child who is still a resident

[b] Includes brothers and Asters of the father, one adopted child, tons and daughters-in-law, and nieces and nephews

to the household headed by her husband's father but, once she and her husband set up their own house, both their incomes go to the newly formed domestic group. Sons contribute to the family income for an average of seven years. They also migrate at around the age of 14 and marry at approximately 21, but they generally bring their wives home to live with their families. Thus, in order to provide the wage income required over a period of twenty years, it can be estimated that the family needs three sons or a combination of four sons and daughters.

This is just enough to ensure the survival and social reproduction of the family. Even this survival is tenuous, as it is subject to adverse events such as a death in the family, a relative drop in wages, the fluctuations of the labour market and other unforeseen circumstances. It must be remembered that these are units of production with practically no savings to fall back on. Consequently, if the household wishes to ensure its reproduction, it must invest.

Given the economic structure of Toxi, the possibilities of investing are extremely limited. The shortage of land makes it difficult to increase production, and the fragmentation of the already small plots means mechanization is not feasible. The lack of a local market for services, handicrafts or cottage industries, along with the monopoly of trade by large-scale intermediaries, means that there are very few opportunities for investing outside agriculture. Since it is extremely difficult to invest in production, the analysis suggests that families invest in biological reproduction.

Data from fieldwork observation showed that during the first two stages of the domestic cycle all extra resources in the family are devoted to the raising and feeding of the largest possible number of children. Hence, the mother never migrates because her constant presence and her continuing pregnancies are needed to produce and raise children. The greater the number of children, the larger will be the eventual surplus from the migratory income contributed by each of them. Table 7.8 makes it clear that the tendency to migrate increases with the number of children.

If it is assumed that, during its critical stages, the household requires the cash income supplied by four children in order to be certain that the family will reproduce itself, the additional income from a fifth or sixth child would represent a surplus for the family in the later stages. More children, then, will mean more potential surplus. While this study cannot prove that families actually have a larger number of children, the analysis does indicate that family units retain offspring for a longer period than in communities like Dotejiare. By having and retaining more children, families get an income through relay migration that may provide a cash surplus for investment.

If they cannot invest in production, given the economic conditions in the village, what do they do with this surplus? The data suggests they invest it in the education of the younger children.

In Mexico, traditionally, the elder brothers and especially elder sisters help pay for the schooling of the younger brothers, and this was very much the case in Toxi. Today, however, with the increasing job opportunities for women, particularly in the nearby electrical-appliance factory, parents are considering their daughters'

Table 7.8 Percentage of members who migrate according to number of resident children in Toxi

Number of resident children	Percentage of migrants over total in each category				
	Father	First child[a]	Second child	Third child	Others[b]
1–3	36	36	14	4	0
4–6	36	36	24	9	6
7–11	35	50	50	7	0

[a] Refers to the oldest child who is still a resident
[b] Includes brothers and sisters of the father, one adopted child, sons and daughters-in-law, and nieces and nephews

Table 7.9 Average length of schooling of children aged 4–13 according to the domestic cycle

Village	Domestic cycle (mother's age)					
	18–25	26–30	31–38	39–44	45–50	51 over
Toxi	–	2.2[a]	4.2	4.4	5.0	7.0
Dotejiare	–	2.0	2.5	3.0	3.0	6.0

[a] The number corresponds to primary-school grades. Cases not included: 3 or 3.1 per cent in each village

schooling more important than they previously did. This applies principally, however, to primary education and only sometimes to secondary-school level.

Table 7.9 shows the levels of schooling of children between the ages of 4 and 13 at the different stages of the domestic cycle.

7.4 Conclusions

Toxi and Dotejiare represent two different moments within a single process of the decomposition of the traditional peasant economy based on domestic production. In Toxi, a high level of unemployment has resulted from a combination of the fragmentation and erosion of the land; the disappearance of traditional occupations, craft production and cottage industries; the growing centralization of commerce and large industries in Mexico City; and the unequal terms of exchange through prices for agricultural products, taxes and other financial channels. As Toxi became progressively integrated into the national market, a relative surplus population that could not find work in the local area was created. This situation has not been relieved by the small amount of jobs created by the increase in services and by a few factories in the region. Since 1950, the excessive centralization of industry has created sources of employment only in Mexico City. According to the analysis, the people from Toxi have sought employment in the city so that Toxi has become a proletarian suburb, no longer a peasant community, since slightly more than half of its workers are wage labourers. If it were located within the limits of the metropolitan zone it would be classified as a proletarian neighbourhood. However, since it is actually located 250 km from the city, the unemployed workers of Toxi have to travel to the city in order to obtain jobs, and consequently, are migrants.

Eighty per cent of the households in Toxi have one or more wage labourers, and of these 86 per cent migrate to work in Mexico City. They are, however, still considered resident members of family groups in Toxi. In the city, most join the ranks of the underemployed and of those with temporary, low-paid jobs of minimum productivity. A greater number of men than women are involved in this type of migration. The women are exclusively employed as paid domestic workers.

The dynamics of this temporary migration can only be understood in relation to the domestic cycles of the families in the villages. In the first two stages, when the mother is between 18 and 30, the father and mother must work intensively, since they carry the whole work-load of the house-hold. The mother is no longer involved in the productive activities that used to be part of the peasant domestic production. She can no longer increase the family income through the production of handicrafts or through small trading. Her energies are thus displaced from production and concentrated on reproduction.

Having many children close together may mean the survival of the family unit, so the mother rarely migrates to work outside the community. She remains in the village working occasionally in local wage labour as a servant or washerwoman, carrying out at the same time her own unpaid domestic chores. She also helps in agriculture when needed, and cares for the livestock when her husband leaves to do migrant work. The intensity of her work does not decrease, as it once did, when her daughters grow up. They can no longer help her with her chores, because they must now go to school and shortly afterwards migrate to work in the city. In short, these economic changes mean that the responsibility for this expanded domestic labour falls, as never before, upon the mother alone.

The father's work intensifies during the first two stages of the domestic cycle as he combines his own agricultural work with migratory labour. But in the third stage the older daughters or sons begin to take his place in migratory work. When the first children begin to separate from the domestic group, after the fourth stage of the domestic cycle when the mother is 40 years old or more, the daughters and sons who follow take their turn in migratory labour. This relay migration represents a strategy to provide a cash income for the household in each stage of the cycle. This extra income is indispensable in financing the family's maize production, which, although it hardly provides food for half a year, enables them to hold on to their land and their way of life in the village.

In order to carry out this strategy to ensure the family's minimal social reproduction, the family needs at least four children. With more than four, the unit can save the extra income supplied by the additional children. This income strengthens the family's possibilities of reproduction and allows it to invest in the education of the younger children. Given the economic conditions of the community, this type of investment is the most practical. These mechanisms explain why parents continue to have large numbers of children even when there is no land to give to them or any employment for them in the village. It also explains why the children stay longer in their father's house in Toxi than in Dotejiare.

The situation in Toxi is in striking contrast to that in Dotejiare, where a peasant agricultural economy still predominates. In Dotejiare, the average number of children living with their families is smaller and there is a greater number of extended families among the wealthier group. The young people are the ones who migrate, especially the younger women, and they do so during the third and fourth stages of the domestic cycle (when the mother is between 31 and 44 years old). However, migration in Dotejiare does not follow a deliberate strategy of the family group. In fact, there is no pattern in out-migration in the village except that the predominantly female migrants are generally engaged in street vending in the city, and rarely send money home.

It may be said that in Dotejiare the greater availability of land and the income from local extractive activities have slowed down the process of rapid impoverishment from which Toxi suffers. The analysis describes a situation in which, at least in this type of community close to an industrial metropolis, a process is taking place that is similar to that of Europe in the last century. Thus, fragmentation of land has brought about the impoverishment of the domestic units and a resulting tendency to invest in their own expanded reproduction.

There are also similarities in that the extended families tend to be those that are still associated with agriculture and have the most resources. A significant difference, however, is that in this study the rich families are not the only ones that follow the strategy of the stem family, of sending their children to live in the city and thus diversifying their contacts with different sectors of the economy. Faced with the impossibility of providing land for their children, families with small landholdings also practise the strategy when they send their older children to the city; temporarily at first but later permanently. This finding supports the hypothesis introduced at the beginning of this chapter that the stem family is not a type of family composition but a strategy and one that expands.

Whether a family will be able to follow the cultural norms of inheritance or not will depend on the surrounding economic conditions. In rural Mexico the traditional custom of ultimogeniture is alternated with the juridical norm of equal inheritance, depending on the availability of land or employment for the older sons. In rare cases in which the family is able to obtain enough land, temporary migrants will return to Toxi to live and work in subsistence agriculture. The majority of them, though, after helping their parents' household during the critical years, are forced to settle permanently in the city. It would seem, then, that this strategy is functional for the rural community but dysfunctional at a global level since these permanent migrants are increasingly unable to find stable employment in the city.

If this is true, it runs contrary to the thesis that industrial capital 'maintains' the subsistence peasant economy in order to economize the social costs of the reproduction of labour. Of course this has been and will continue to be one of the functions of the peasant economy in industrial development. However, the analysis in this chapter indicates that the dynamics of the social reproduction of labour in rural areas is beyond the control of the capitalist sector. The decision to use their children's wage labour to subsidize a declining peasant economy has direct consequences on overall development. It can no longer be in the interest of the capitalist sector to maintain agricultural enclaves that send thousands of migrants to industrial centres where there is no possibility of their being employed. This would explain why capital has been particularly interested in birth control and in modernizing peasant communities.

The strategy of relay migration adopted by peasant families is one factor that helps explain the persistence of this inefficient subsistence agriculture, in spite of dramatically unequal terms of exchange with the industrial sector. It also helps explain the indifference and the opposition of peasants to birth-control programmes and the cultural emphasis on the role of women primarily as reproducers.

This does not mean that peasants deliberately and consciously decide to have more children. In spite of the decline in mortality that allows a higher survival rate for children and a longer life-expectancy for parent, the latter face no urgent or immediate reason to limit the number of births. On the contrary, in terms of family income the traditional large family has found a new function for children, which ensures the social reproduction of the household.

To summarize, this chapter contends that relay migration, among other factors, helps explain the short-term persistence of the peasant economy that enables peasants to resist the pressures that lead them to become wage labourers. With no economic or ideological reasons to reduce the birth rate, but rather with benefits to be drawn from it, rapid population growth has continued in these villages.

Faced with integration into the national market on very unequal terms, the peasant families of Toxi have not suffered progressive deprivation passively. If the income sources that were once available in the rural village are now concentrated in the city, they are sending 'branches' to these activities. Through relay migration, they bring back to the rural household part of the wealth that the city extracts from them through a process of unequal development. The city controls the circulation of capital goods and money; the peasants, holding on to their ever-shrinking land plots, control the circulation of their only 'capital': their children.

References

Arizpe, Lourdes, 1972: *Parentesco y economía en una localidad nahua* (Mexico City: Instituto Nacional Indigenista).

Arizpe, Lourdes, 1978: *Migración, etnicismo y cambio económico* (Mexico City: El Colegio de México).

Balán, Jorge; Browning, Harley L.; Jelin, Elizabeth, 1973: *Men in a Developing Society: Geographical and Social Mobility in Monterrey, Mexico* (Austin: University of Texas Press).

Berkner, Lutz, 1972: "The Stem Family and the Development Cycle of the Peasant Household: An 18th-century Austrian Example", in: *American Historical Review*, 77, 2 (April): 398–418.

Castells, Manuel, 1975: "Immigrant Workers and Class Struggles in Advanced Capitalism: The Western European Experience", in: *Politics and Society, 5 (March):* 33–66.

CLACSO (Consejo Latinoamericano de Ciencias Sociales) 1973: *Migración y desarrollo* (Buenos Aires: CLACSO).

CLACSO (Consejo Latinoamericano de Ciencias Sociales) 1975: *Migración y desarrollo* (Buenos Aires: CLACSO).

Chayanov, Alexander V. (edited by Thorne, D.; Smith, R.E.; Kerblay, B.), 1966: *The Theory of Peasant Economy* (London: Irwin).

Colegio de México, 1970: *Dinámica de la población en México* (Mexico City: Centro de Estudios Económicos y Demográficos).

Deere, Carmen Diana, 1978: *"The Differentiation of the Peasantry and the Economic Basis of Family Structure: A Peruvian Case Study"* (Mimeo).

Díaz Polanco, Héctor, 1974: "La economía campesina y el impacto capitalista" (Mimeo).

Elizaga, Juan Carlos, 1970: *Migración a las áreas metropolitanas de América Latina* (Santiago de Chile: CELADE).

Goody, Jack, 1968: *The Development Cycle of the Domestic Groups* (Cambridge: Cambridge University Press).

Habakkuk, H. John, 1960: "Family Structure and Economic Change in Nineteenth-Century Europe", in: Bell, Norman; Vogel, Ezra (Eds.): A Modern Introduction to the Family (Glencoe-New York: The Free Press): 538-543.

LePlay, Frédéric, 1871: *L'Organisation de la famille, selon le vrai modèle signalée per l'histoire de toutes les races et toutes les temps* (Tours: Mame et Fils).

Meillassoux, Claude, 1977: Mujeres, graneros y capitales (Mexico City: Siglo XXI).

Muñoz, Humberto; De Oliveira, Orlandina; Stern, Claudio, 1977: Migración y desigualdad social (Mexico City: El Colegio de México-UNAM).

Singer, Paul, 1975: *Población y desarrollo* (Mexico City: Siglo XXI).

Tilly, Louise; Scott, Joan Wallach, 1978: *Women, Work and Family* (New York, Holt, Rinehart & Winston).

Young, Kate, 1978: "Economía campesina, grupo doméstico y migración", in: *América Indígena, 38, 2* (June): 279–302.

Chapter 8
Agrarian Change and Women's Rural Out-migration in Latin America

"Females are more migratory than males" wrote Professor E. G. Ravenstein in 1885, in his pioneering attempt at discovering the laws of migration (Ravenstein 1885). Today, one hundred years later, while increasingly sceptical that such historical 'laws' may exist for all societies and all times, we are still facing the task of explaining why, indeed, "females are more migratory" in most Third World countries at present, especially in certain contexts of agrarian change.[1]

In a comprehensive review of data on women's migration in the Third World, Youssef, Buvinic and Kudat (1979) have advanced our understanding by showing that gender differences do create distinct patterns of migration and by identifying some regularities in women's migration among different countries. Previously, more formalized models, such as the Todaro-Thadani model, in which income differentials were used as the main factor to explain migration, two additional factors were brought in, the desire for social mobility and the wish to find a husband, in order to try and predict female migration (Todaro/Thadani n/a; Harris/Todaro 1989).

However, this article contends that, in order to explain female patterns of migration, women must be understood within the context of the dynamics of social structure in the sending areas. It is this context which ultimately determines who will be most likely to migrate among the women and men of the family or the community. In what follows, an attempt is made to explain why certain women of a given age, marital status or social and economic background, are the most prone to migrate out of specific rural communities in Latin America.

---

[1] This text was first published as: "Relay Migration and the Survival of the Peasant Household", in: Jorge Balan (Ed.): *Why People Move* (Paris: UNESCO, 1982): 187–210 and it was republished in Safa, H. (Ed.): *Third World Urbanization* (New Delhi: Oxford University Press, 182): 187–210. This text is used with permission of UNESCO and the permission was granted on 4 July 2013 by Isabelle Nonain-Semelin on behalf of UNESCO.

L. Arizpe, *Lourdes Arizpe*, SpringerBriefs on Pioneers in Science and Practice 10, DOI: 10.1007/978-3-319-01896-6_8,

8.1 Peasant Corporate Communities

Peasant communities, assuming no demographic pressure on the land (caused by either concentration of lands or natural population growth), tend to seek a balance between their populations and their systems of production (Butterworth 1982). In many cases, especially in long-established communities with highly developed Indian peasant traditions, social and economic mechanisms are generated which tend to redistribute the economic surplus and the population imbalances among the community households (Arizpe 1982). Mexican peasants, for example, have developed a highly complex hierarchical system that, among other things, redistributes the surplus of individual households within the community through ritual expenditure, free food, and credit availability for poorer farmers. At the same time, kinship, ritual kinship and marriage ties allow a redistribution of people from poorer households to those with greater resources, for example, through uxorilocal marriage or the adoption patterns of ritual kinship. As a result, women of all ages have had a place in some household where, as is typical of a peasant economy, they can contribute to the work of the family unit (Arizpe 1978). Such institutions function as a social security system whereby the kin group, or the community, is responsible for the welfare of all of its members. Widows, orphaned girls, unmarried women and even unwed mothers normally belong to some household, and are able to eke out a living in the rural community.

An important point is that mechanisms like these function independently of economic levels, and therefore the departure of women migrants cannot be predicted by correlation analyses of female migration and objective criteria such as income levels, unemployment, capital accumulation, etc. This does not necessarily mean that women in peasant corporate communities had egalitarian status or led fuller lives, and poor young girls taken in by richer kinfolk became servants more often than not. In fact, the old word for servant in Spanish, '*criada*', literally means someone who has been brought up in the household. Nor did the protection of widows or unmarried women ensure a good life. But women did not usually find themselves unable to live or work in their rural communities.

The transition to a permanent outflow of women from peasant communities begins with the uneven internal development at the national level. As peasant communities become monetized, they begin to lose their surplus through price differentials between agricultural products and manufactured goods, and the loss of their traditional income-generating activities, many of them carried out primarily by women. This is one of the factors that make women more 'available' for migration.

The moral and religious constraints to women's mobility are broken when peasant families feel an urgent need for additional incomes, or find that they can no longer feed their offspring. It is at this point that patterns of migrant selectivity are generated, on the basis of the ideological norms on which gender differences are established in the division of labour and in social behaviour. Generally, in Latin America, because girls are normally not fully involved in traditional agriculture, and because they have better employment opportunities outside, they are the first to be sent out as migrant labourers.

But this push factor would not have such a strong effect had there not been an equally strong pull factor. The ideological constraints which restrict female geographical mobility are overcome because an 'honourable' labour migration path is found in employment in urban domestic service (Jelin 1978; Grau 1982). Young girls are entrusted to the care and protection of the urban matron for whom they work as servants. Such employment is plentiful, since a large demand for women servants is generated through the growth of Latin American cities where the upper and middle classes can afford to pay for such services (Arizpe 1978b).

The initial flow of women towards the cities grew larger when the peasant economy became gradually undermined by the extraction of surplus to finance industrial and urban growth, and segments of the rural economy became labour reserves for commercial agriculture and urban industry.

8.2 Peasant Communities as a Labour Reserve

Historically, peasant communities in Latin America have provided cheap labour for semi-feudal, semi-capitalist haciendas, and for plantations having direct foreign investment (CLACSO 1973). This labour has been predominantly male, although men frequently took their wives and children along to help in cash cropping. In fact, wages were often so low that the only way to eke out a living was by involving the whole family. Women, however, rarely migrated on their own to work as agricultural wage labourers.

The process whereby formerly independent subsistence communities became labour reserves was intensified with the development of agrarian capitalism in rural areas through the monetization of the peasant economy (Fei/Ranis 1964; CLACSO 1975). As land, labour and inputs are increasingly exchanged on a cash basis rather than through the traditional exchange systems based on complex kinship, lineage, residential and community ties, the cash needs of peasant families increase (Klein n/a). Draught animals, ploughs, and seeds have to be paid for, and, with the modernization of agriculture as well as soil erosion due to intensive cultivation, fertilizers and insecticides become indispensable. Other modern services such as medical care, transport, drinking water, schooling and electricity must also be paid for, and, at the same time, the mass media encourage the purchase of consumer goods. While their cash needs soar, however, their profits from agriculture shrink with the relative decline in agricultural prices, as do their profits from traditional crafts, trade, and cottage industries.

With an ever-increasing deficit, peasant families face the choice of abandoning their small plots of land to become wage labourers, or staying on by sending out migrants to earn cash incomes (Mayer 1961; Elizaga 1970). The different strategies they adopt to diversify their sources of income are governed by the possibilities open to them in their immediate surroundings (Peek 1982). These may be local wage-labour in agriculture or in the towns, or internal migrant wage-labour in agriculture or in cities, or international migrant labour (Roldán 1982; Spindel 1982). Or they may combine seasonal with temporary (this can mean several years) or permanent migrant labour (Arizpe 1978).

If we analyse the various options, it is interesting to note that the one which offers the most advantage is migration of young women to domestic service in the cities (Smith 1973). Their labour is needed in agriculture and at important times, as during the harvesting season, they may ask for permission from their employers to go home. Since they live in their employer's house, their expenses are minimal, and their earnings can be high. Furthermore, the father has greater control over them than he would have over his migrant sons.

One aspect that is rarely touched upon is the effect of monetization on the older women of the household. Old women, grandmothers or aunts, can no longer be supported and are expected to find their own sources of income. Since their labour is no longer needed for traditional crafts and cottage industries, and since small-scale trade is pushed aside by large-scale oligopolistic trading, they often have to resort to begging. In the best of cases, many of them go to the cities to live with migrant offspring who may be able to support them.

When the need for a cash income becomes permanent in the peasant household, a strategy of relay migration may be adopted in which the father, daughters, and sons take turns at migrant labour (see Chap. 7 in Balán 1982; Arizpe 1982). During the early stages of the domestic cycle, the father migrates seasonally. When the eldest offspring is old enough, he or she is sent temporarily or permanently to the city; other sons, but daughters especially, follow as the eldest migrants get married or lose their bond with the rural household. Interestingly, in the late stages of the domestic cycle, the remittances of the younger daughters are used to finance the education of the youngest sons in an attempt to increase the household's chances for a higher income in the future.

Regularities in migration strategies are reflected in certain patterns of age and out-migration in women from these communities. Overall, women tend to leave between the ages of 18 and 20, and then between 22 and 25; there is only a trickle of migrants after that age, until the age of 40 when again there is a concentration between 40 and 45 (Arizpe 1982). After that, there is only a trickle again. This pattern indicates the weight of the marriage factor in women's out-migration. The largest group of women tend to leave at a marriageable age (18–20). Of those who stay, any who are still unmarried by the time most young marriages in the village have taken place leave then (22–25). After that, most migrants either leave with their husbands or because of a marital break-up. At the age of 40, widows or old couples may decide to move to the city to live with offspring. Beyond that, elderly women will migrate only out of desperate necessity or widowhood (Arizpe 1982).

Important as the marriage factor is in female migration, the weight of social class also becomes evident when we compare ages with the kind of household the migrant comes from. Again, regularities emerge. The relay-migration strategy follows a distinct pattern in peasant households: the eldest daughters migrate at a significantly early age, sometimes 10 or 12 years, or, more usually, between 13 and 15, a fact which indicates the urgent need for a migrant income in the early middle stages of the domestic cycle (Arizpe 1987, 1982, 1981). Daughters who are in the middle of the birth order migrate at a slightly older age, between 15 and 17, indicating that their elder sisters are still sending remittances. There is no age regularity in the out-migration of the youngest daughters, indicating that their cash income is no longer urgently needed to finance cultivation, although it becomes important for investment, i.e. in the youngest son's education.

Patterns are not so regular in other types of households. Where the main family income comes from craft production or some cottage industry, female out-migration is slightly lower, and ages at the time of leaving are higher—presumably reflecting the fact that their labour is needed in the family enterprise. Women's out-migration is highest in households where the father or mother is self-employed in petty trade or the services. Migration occurs at all ages, but especially at early ages. In such cases, it would seem that remittance to the family is not counted on. Rather the young daughter is expected to fend for herself since her parents' household has little to offer for the future.

There is little to add to the truism that the greater the number of offspring, the greater the propensity of women to migrate. However, it is important that social class may offset this tendency since, as is well known, the richer peasant and farming household is able to support, and needs, an extended family system for its enterprise. Thus, if there is female out-migration from such households, it is usually due to high social or educational aspirations for a young woman.

In analysing the context of women's situations and determining which women migrate from which types of households and communities, it is also important to observe which women stay. Although this phenomenon cannot be analysed here at length, it is worth noting what happens to a woman who rarely migrates—the mother in the peasant household. Her work-load increases dramatically but, once again, invisibly. Once her children begin to attend school, and other females have to leave the household, she must substitute for all of them—fetching water and wood, caring for draught animals and fowl, seeding, weeding and harvesting, feeding male workers in the fields, carrying the crop back to the house, collecting and processing food—while still carrying her regular load of domestic work, cooking, caring for children and husband, washing, etc. The only event which makes her additional work-load visible is the migration of her husband, when she has to run the farming unit. It is also important to note that her presence is central to the peasant household. As long as the mother does not migrate, the family has a homestead; when even she migrates, the family has become totally proletarianized, and its members are then individual, independent migrants.

Another important aspect that needs to be mentioned is that once the rural area becomes a labour reserve, the higher the number of offspring, the better the chances of the peasant family's survival in exploitative conditions. As I have argued elsewhere, parents do not decide to have more children so as to send them out as migratory labourers; the fact is simply that, given the advantage that a large family provides, parents see no reason to stop having children. In this sense female out-migration to the cities, particularly to domestic service, is a structural phenomenon in that it is inherently linked to the monetization of the peasant economy as well as to its main population trend.

8.3 Family Producers for the Agricultural Market

The patterns of female out-migration are different for farming households, both low-income and highly capitalized, which produce a cash crop for the market. Among low-income households, the main difference arises out of the labour requirement for cultivation. When crops are labour-intensive—flowers, strawberries, or peanuts, for example—and the family unit cannot afford to pay labourers, young women are kept at home. Female labour is crucial to the livelihood of such households since their labour is needed to increase productivity to compete in the market, frequently against large capitalist enterprises. Out-migration of daughters is discouraged, since cash remittances could not offset the cost of hiring labourers (Spindel 1982).

But it is worth noting that this situation is the only one among all those analysed in which a disincentive for women's migration is strong enough to balance the pull factor of mass media and educational programmes. In a household which runs a full-scale capitalist enterprise with high technology, there are no push factors encouraging women to migrate; but higher expectations of educational and social mobility may encourage migration.

8.4 Agricultural Wage Labourers

The loss of resources through uneven exchange with the urban industrial sector has made small-scale agriculture no longer viable in many parts of Latin America. For many peasant households this has meant the loss of their small plot of land. Once the family becomes proletarianized, there is a pattern of female out-migration, but it is certainly not as regular as that of other types of households. To begin with, following a historical trend, families of wage labourers tend to have fewer offspring. If the family lives permanently in a village and, say, only the father and the elder sons migrate on a seasonal or temporary basis, young girls tend to stay at home and migrate only after about the age of 18.

If the family moves about in an annual cycle of agricultural wage labour, however, the sons and daughters tend to lead separate lives early on. The father has little control over their labour, unless they all work as a unit on a piece-work basis. But normally, as soon as they are adolescents, young women and men can command a wage of their own. The young woman's decision to migrate independently depends on personal factors, and her destination is strongly influenced by the availability of jobs in agriculture, agro-industry or the cities.

The recent trend of foreign investors in local export crops and agro-industry to give preference to young rural women is changing the patterns of female out-migration. Women who would otherwise have migrated to the city may stay in the region or may migrate to another rural area, thus increasing rural–rural migration. Young women are also being drawn into agro-industry.

The trend is reinforced by redeployed industries which settle in rural areas and employ young women for textile, clothing, electronic or electrical assembly work, frequently on a piece-work basis. The big question here is whether or not such employment will be permanent. If it is, voluntary turnover of young rural women will lower migration to the cities. If it is not, and the industries use female labour power while it is cheap and leave when it becomes more costly, older women, many of them heads of households will be left uprooted, unemployed, unattached, and with poor prospects of survival. In that case, migration to the cities will have been delayed for some years, but these women will inevitably end up in the impoverished shantytowns of cities which offer them no future (Orlansky/Dubrovsky 1978).

8.5 Conclusion

This chapter has looked at the structural constraints and pressures of women's out-migration from rural communities in Latin America. This process is inherently linked to changes in agrarian systems and to employment opportunities brought about by the uneven development of the traditional rural economy as opposed to the modern urban sector.

In the author's view, the structural dimension of female migration must first be broadly outlined in order to understand the position of women in the dynamics of the development processes which shape large-scale internal migrations. These are the conditions necessary for female migration, but they are not sufficient to explain the complex patterns of migration which have been traced here. In contrast to male migration, it is quite clear that marriage and life-cycle factors play an important part in women's decisions to migrate.

These factors have only been mentioned briefly, but they are crucial to the mode of analysis that is proposed here. Life-cycle considerations, in so far as they are reflected in the social organization, form an intermediate structure between macro-economic pressures and women's individual lives, and channel their migration.

But marriage and life-cycle factors must be broken down into the various norms applied to women's family obligations, sexual restrictions, social reputation, ideological control, marriage contracts, and so forth. Marriage norms, for example, cannot simply be added to economic factors to explain patterns of female out-migration, because the opportunities for rural employment are largely determined by sexual, social and marriage constraints. Thus, trends in migration can be explained only with reference to both economic and social factors. For example, female migration for domestic service is both an economic strategy of a peasant household and a social strategy that preserves the 'femininity', i.e. marriageability, of the young migrant.

The other dimension that, must be taken into account, once the channelling factors of development processes and of social norms for women are established, is what kinds of decisions women are able to take when faced with these options. This dimension has not been touched upon here, mainly because so little is known in this area, but it must certainly be incorporated into models of female migration.

Finally, in economically dependent areas, it is important to examine female rural out-migration in relation to development policies. State policies towards the agrarian sector can have unexpected results when the sexual division of labour in the rural household is ignored. For example, men are expected to migrate to the urban industries in surplus labour models, but it is women who have predominated in rural–urban migration. Among other factors, this contributed to the well-known expansion of the service sector (tertiarization) and the informal sector. These distortions of the development process are, indeed, the result of policies of industrialization and agrarian change, but as the analysis of female migration reveals, they are also an outgrowth of the expectations of women and the restrictions imposed upon them.

References

Arizpe, Lourdes,1978: Migración, etnicismo y cambio económico (Mexico City: El Colegio de México).

Arizpe, Lourdes, 1978b: "Mujeres migrantes y economía campesina: análisis de una cohorte migratoria a la ciudad de México, 1940-1970", in: América Indígena, 38: 303–325.

Arizpe, Lourdes, 1982: "Relay Migration and the Survival of the Peasant Household", in: Balán, Jorge (Ed.): Why People Move (Paris: UNESCO): 187–210.

Arizpe, Lourdes; Aranda, Josefina, 1981: "The Comparative Advantages of Women's Disadvantages: Women Workers in the Strawberry Export Agribusiness in Mexico", in: Signs, 7,2 (Winter): 453–474.

Butterworth,Douglas, 1982: "Migración Rural-Urbana en América Latina: es estado de nuestro conocimiento", in: América Indígena, 21,1: 52–85.

CLACSO (Consejo Latinoamericano de Ciencias Sociales) 1973: Migración y desarrollo (Buenos Aires: CLACSO).

CLACSO (Consejo Latinoamericano de Ciencias Sociales) 1975: Migración y desarrollo (Buenos Aires: CLACSO).

Balán, Jorge (Ed.), 1982: Why PeopleMove (Paris: UNESCO).

Elizaga, Juan C., 1970: Migración a las Áreas Metropolitanas de América Latina (Santiago de Chile: Centro Latinoamericano de Demografía).
Fei, John C.H.; Ranis, Gustav, 1964: Development of the Labor Surplus Economy: Theory and Policy (Homewood, Illinois: Richard A. Irwin Inc.).
Grau, Ilda Elena, 1982: "Trabajo y vida cotidiana de empleadas domésticas en la Ciudad de México: Un estudio cualitativo", in: León, Magdalena (Ed.): Sociedad, Subordinación y Feminismo (Bogota: ACEP): 167–187.
Harris, John R.; Todaro, Michael,1989: "Migration, Unemployment and Development: A Two Sector Analysis", in: American Economic Review, 60,1 (March): 126–142.
Jelin, Elizabeth, 1978: "Migration and Labor Force Participation of Latin American Women: The Domestic Servants in the Cities", in: Signs, 3,1 (Winter): 129–141.
Klein, E., n/a: "Peasant Systems, Employment and Agrarian Change in Lain America, Centre for Latin American Studies, Cambridge University (Mimeo).
Mayer, Philip, 1961: Townsmen or Tribesmen (Kampala: East African Institute of Social Research).
Orlansky, Dora; Dubrovsky, Silvia, 1978: "The Effects of Rural-Urban Migration on Women's Role and Status in Latin America", in: Reports and Papers in the Social Sciences (Paris: UNESCO).
Peek, Peter, 1982: "Agrarian Change and Rural Emigration", in: Balán, Jorge (Ed.): Why People Move (Paris: UNESCO): 59–71.
Ravenstein, Ernst Georg, 1885: "The Laws of Migration", in: Journal of the Statistical Society, 48,2 (June): 167–235.
Roldán, Martha, 1982: "Subordinación genérica y proletarización rural: un estudio de caso en el Noreste Mexicano", in: León, Magdalena (Ed.): Las Trabajadoras del Agro II (Bogota: ACEP): 75–102.
Smith, Margo L., 1973: "Domestic Service as a Channel of Upward Mobility for the Lower-Class Woman: The Lima Case", in: Pescatello, Ann (Ed.): Female and Male in Latin America (Pittsburgh: University of Pittsburgh Press): 191–207.
Spindel, Cheywa, 1982: "Capital, familia y mujer: la evolución de la producción rural de base familiar, un caso en Brasil", in: León, Magdalena (Ed.): Las Trabajadoras del Agro II (Bogota: ACEP): 227–247.
Todaro, Michael P.; Thadani, Veena N., n/a: "Towards a Model of Female Migration" (Mimeo).
Youssef, Nadia; Buvinic, Mayra; Kudat, Ayse (Eds.), 1979: Women in Migration: A Third World Focus (Washington: Office of Women in Development).

Lourdes Arizpe interviewing women migrants near the Merced market in Mexico City in 1972

Chapter 9
Women in the Informal Labour Sector: The Case of Mexico City

In recent years, the informal labour sector has increasingly represented a testing point for theories of development.[1] The proliferation of informal jobs in developing countries has been considered alternately a stage in the process of development and a blind alley leading a country back into underdevelopment. But social scientists and policymakers have rarely recognized that the majority of those left out of the formal occupational structure are women. It is, however, very difficult to establish the heuristic boundaries of the informal labour sector, particularly with respect to women. Are we referring to the intermittent part-time activities of women outside the household both in cities and in rural areas? But men also engage in such activities, for example, as street peddlers. Is the unpaid work of the wife and young unmarried daughters in a family enterprise such as a store an informal job? If unpaid labour is to be included in the informal labour sector, then women's voluntary community service and their unpaid domestic labour must also be taken into account. Moreover, since informal labour also comprises work not regulated through a contract, all low income, non-contractual jobs registered as formal occupations, such as paid domestic service, belong to this classification. Many low income and low productivity jobs included in the formal occupational structure and registered in national censuses, even when such a contract does exist, can be considered as a continuation of informal jobs as well and thus must be analysed within the informal labour sector.

This paper assumes that the nature of the informal labour sector in a developing economy is a direct outgrowth of the type of industrialization a country is undergoing. Within this framework, this paper explores the degree of occupational choice that women have within the structural margins of employment. In recent years, the question of choice has become central to an analysis of women's participation in economic and political activities. Noting a decline in women's economic participation as development proceeds in certain underdeveloped countries,

[1] This text was first published in: *Signs* (Chicago: The University of Chicago Press), 3,11 (Autumn 1977): 25–37. The permission to republish this was granted on 12 June 2013 by Perry Cartwright for the University of Chicago Press with grant number 107279 (Ref. No. 0049537789).

L. Arizpe, *Lourdes Arizpe*, SpringerBriefs on Pioneers in Science and Practice 10,
DOI: 10.1007/978-3-319-01896-6_9, © The Author(s) 2014

Ester Boserup asks whether this is due to difficulties in finding jobs or to voluntary withdrawal because of family obligations. A definite answer is perhaps empirically unverifiable, because of the ideological nature of assumptions on which decisions are made. However, understanding the nature of the occupational options open to women will clarify the issues.

9.1 The Nature of the Informal Labour Sector

An individualistic theoretical viewpoint would assume that the types of jobs taken, the sex distribution, and the size of the informal labour sector are a result of random individual decisions. Research based on such a viewpoint would centre on the way in which women weigh their options and decide on a course of action, but could not explain why those options alone were available. The usefulness of this individualistic approach in studying the informal labour sector is therefore quite limited. It can help show why certain women choose particular kinds of jobs at certain times, but not why women as a group tend to enter the informal labour sector rather than formal employment. The answer can only be found in understanding how capitalistic development affects women's economic roles.

According to developmental theory, labour shifts from agriculture to the modern sector and is absorbed by manufacturing industries, following the pattern of industrial growth in developed countries.[2] Although the informal labour sector is not explicitly referred to, we must infer that it is regarded only as a bridge over which workers pass in shifting from one sector to another. In most Latin American countries, as well as those in Africa and Asia, however, data show that the displaced labour from agriculture enters the informal labour sector, most often in the cities which provide a large market for such jobs, and remains there (Stavenhagen 1972; Cardoso/Faletto 1973; Frank 1972). "The excess population (which in a system operated by peasants lives on the family farm) shows itself… mainly in domestic service, trading, and casual jobs" (Myrdal 1968: 204). Whereas these informal activities constituted a secondary source of income for the peasant household, they become a primary source of income as agriculture is increasingly dominated by wage labour, and people become wholly dependent upon them economically. At best, workers go into the service sector rather than manufacturing. Even in countries with moderate industrial growth, such as Brazil and Mexico, the jobs created by industrialization have been unable to compensate for the loss of employment in agriculture (Singer 1975; Suárez 1973). Rapid demographic growth has made this situation even more acute.

---

[2] The main exponent for this theory is Walter Whitman Rostow. See: Rostow, Walter Whitman, 1960: *The Stages of Economic Growth: A Non-Communist Manifesto* (Cambridge: Cambridge University Press).

Photo: The 'Marias', Mazahua women migrants in Mexico City with Lourdes Arizpe (1973)

An alternative theoretical explanation to the expansion of the informal labour market in Latin America is provided by the concept of marginality. According to Quijano (1970) and Nun (1969), the population displaced from agriculture has been unable to find a place in the modern sectors of developing economies because of the nature of their industrial growth. Whereas the industrial reserve army in central capitalist economies provides an accommodating supply of labour, in underdeveloped countries labour absorption in industrial employment has not expanded at an adequate rate in proportion to the increase of available labour, so that large numbers of workers have no possibility of finding formal employment. This marginal population, waiting to enter the formal occupational sector, survives by low income, intermittent wage or self-employment. In other words, workers are pushed into the informal labour sector and into the services as a result of insufficient demand in the manufacturing sector, and they are destined never to leave them.

Recent studies, however, tend to cast doubts on some aspects of this theory. There has been a slow but steady absorption of workers, even unskilled ones such as rural migrants, into the formal sector in Brazil, Argentina, and Mexico, which have a moderate industrial growth rate (de Olivera 1976). Moreover, it has been argued by Hobsbawm (1975, 1982) that marginality has always existed in capitalist economies and that the difference between the industrial reserve army and the marginal population is not qualitative but quantitative. It has been suggested that 'marginals' do fulfil specific functions within the economy. In Brazil, for example, full- employment policies have not prevented street vendors from pursuing their activities.[3] In Mexico City, certain manufacturers, who could not place their products in supermarkets because of monopolistic trade practices, used street vendors successfully to increase sales. Finally, participation in the informal labour sector does not automatically result from lack of jobs in the other sectors. Some informal activities are traditional in an urban society and usually fulfil specific functions. Indeed, participation in some traditional urban informal jobs seems to be

[3] Faria Vilmar, personal communication, February 1976.

unaffected by fluctuations in labour demand. Although increase in surplus labour adds a greater number of workers to those traditional informal activities, it also creates new ones.

9.2 Women in Development: Their Participation in the Informal Labour Market

Statistically, the distribution of women in the labour force does tend to reflect the level of development of a country (Tienda 1974; Boserup 1973). A detailed analysis of female labour participation according to the level of development of the Mexican states showed that "for every unit increase in the index of economic development, there is a corresponding increase of about 3 % points in the aggregate level of women's labour force activity" (Tienda 1974: 15). However, the overall rate of participation can vary simply because of differences in statistical definitions. As Boserup has observed, "Official statistics in a developing country may show female activity rates increasing or decreasing over time without any real difference in the work performed by women being involved" (Boserup 1973: 388). Some countries take women's labour in agriculture into account, while others do not. Usually neither part-time work outside the home nor work for a family enterprise is included in national censuses. The Mexican census, for example, only records women's primary activity, and since this is assumed to be domestic work, the frequent, and in many cases constant, involvement of women in the informal labour section does not appear in official statistics. Since women's domestic work is not classified as an economic activity, 79.9 % of women in Mexico thus appear to be 'economically inactive'.

By and large, women's productive activities decline when they are no longer involved in the agricultural tasks and the cottage industries of a peasant economy. This is particularly evident where rural–urban migration is involved. By the same token, the introduction of new technology, both in the agricultural sector and in industries, displaces women and restricts their access to new jobs (Chaney/ Schmink 1975). During the first stage of industrialization, the types of industries that are established, mainly textiles and leather, are labour intensive and provide additional employment for women. But when mechanization advances, employment goes predominantly to men. For example, whereas women in 1900 comprised 45.3 % of the Brazilian labour force, by 1970, notwithstanding industrial expansion, women represented 21 % of the labour force (Saffioti 1975). Industrial growth, in fact, intensifies sectorial sex inequality in both developed and developing economies. As Schmink has remarked, "occupational segregation has, if anything, shown a tendency to increase in Venezuela" (1974: 15). Cultural factors must then be taken into account to explain differential sex distribution in the economy, especially the sex labelling of jobs and women's attitudes toward work (Schmink 1974: 18; Youssef 1971; Boserup 1973: 387). More specifically, age and marital status influence the possibility of formal or informal employment for women. In Mexico, for example, "the level of participation of single females does seem to vary systematically with development levels, while the rates for ever-married females remains relatively unaltered by degree of development" (Tienda 1974: 16). This may mean that divorced, widowed, or separated women work whether employment is expanding or not, while single women stay at home until industry offers them adequate jobs. Thus the increase in jobs does not benefit the most needy women who support their children, but the young women who will normally leave the labour force when they get married. In Latin America women's involvement in the labour force declines steadily after the age of twenty-five. Is this due to voluntary withdrawal, or to the unavailability of jobs for older women? I suggest that in Latin America, and perhaps in other developing regions, women's participation in formal employment declines with age while it increases in informal activities.

9.3 Women in the Mexican Labour Force

Women represent 20.6 % of the Mexican labour force (Ruíz 1975: 76). This is a low percentage compared to that for other developing nations, which averages 26 % (ILO 1975). This low figure stems partly from the inadequate registration of women's labour in rural areas. Because most of the women are engaged in agricultural tasks intermittently, particularly during seeding, weeding, and harvest time, they are classed only as housewives and are considered 'economically inactive'. However, if we add the 'economically active' women to the 'economically inactive' housewives, 94.7 % of the female population over the age of twelve is involved in work, as compared to 89.7 % of the male population. Significantly, only 17.5 % of the women receive a wage for their labour, as compared to 90.3 % of the men (Ruíz 1975: 134). Keeping in mind the inaccuracy of figures for agriculture, the sectorial sex distribution in the Mexican economy is shown in Table 9.1. Women are overwhelmingly employed in the tertiary sector of the Mexican economy, and about 40 % are engaged in jobs of extremely low productivity and income (González 1975: 120). In the manufacturing sector, they are concentrated in dress manufacturing, food, electrical equipment, and textile industries, primarily as administrative personnel (Ruíz 1975: 130).[4] Women rarely hold high administrative or professional posts; 60 % are workers and employees, as compared to 37.1 % of the men. Status differences between men and women are even more conspicuous in the modern sector of the economy; in Mexico City 25.1 % of male office workers are classified as professional or technical in comparison to 14.6 % of female office workers.

---

[4] I am indebted to Ruíz Harrell for the remaining statistics in this section.

Table 9.1 Sexual distribution in Mexican economy. *Source* Ruíz (1975: 134)

Sector	Women (%)	Men (%)
Agriculture	10.0	47.2
Manufacturing	16.9	16.8
Services	64.5	26.8
Sector not specified	2.5	9.5
Insufficiently specified	6.1	2.7
Total	100	100

Table 9.2 Education and unemployment. *Source* Ruíz (1975: 134)

Educational level	Unemployed Women (%)	Unemployed Men (%)
No schooling	11.0	24.2
Primary school:		
Grades 1–3	31.4	36.3
Grades 4–6	33.5	26.8
Grade 6	12.1	6.4
Secondary school	5.5	4.2
Middle education	4.8	1.2
Professional	1.7	0.9

Sharp differences exist between the male and female unemployment rates in Mexico. In recent decades, women's unemployment has risen at a faster rate (14 %) than that of men (5.7 %). By 1980, in Ruíz Harrel's estimation, female unemployment will overtake male unemployment, and 60 % of all unemployed will be women. Women have greater difficulties than men in finding a job for the first time, particularly in the 30–39 age group, and according to the national census women take a longer time to find a job at all ages. As Table 9.2 suggests, this situation persists even though women seeking employment have more years of education. Although educational background does generally correlate with unemployment, women with the same level of schooling as men have higher unemployment rates.

9.4 Women in the Labour Force in Mexico City

More women work in Mexico City (29.7 %) than in the rest of the country (17.8 %; Suárez 1973: 406). Women's participation in manufacturing has increased since the 1940s with the expansion of industries. In fact, they have been employed at a faster rate than men, who nonetheless hold double the number of middle-professional and technical jobs. In the 1960s, however, the rate of labour absorption in manufacturing decreased in Mexico City, while the service sector expanded, particularly in self-employment for craft production. As a result, unemployment increased in Mexico City during the sixties, particularly for women; female unemployment is twice as large in Mexico City as in the rest of the country (46 % vs. 20 %; Suárez 1973: 387). Suárez Contreras has concluded that urban women, who have educational levels that are 20 % higher than in the rest of the country, refuse to accept jobs of inferior status. On the other hand, in Mexico City, while the rate of female unemployment has increased, the rate of female participation in domestic labour has decreased between 1950 and 1970. In the rest of the country, significantly more women are now in domestic work per family than in 1950. This may be due to the fact that rural women who can find no work stay in the house to help with domestic tasks, whereas Mexico City women in the same situation consider themselves unemployed.

In Mexico City, 72.2 % of working women as compared to 53.9 % of men received less than the official minimum salary in 1970 (González 1975: 120). In the services, which employ 42.9 % of the total female labour force, the figure is 72.9 %. Of the total salaried labour force between 21 and 60 years of age in Mexico City, 18.1 % of the men are in these low-income jobs, as compared to 35.6 % of the women (Muñoz/de Oliveira/Stern 1972). Table 9.3 shows the sex composition by occupation. It is these jobs, especially those of street peddler and unskilled worker in the services, that must be closely examined in relation to the informal employment of women.

9.5 The Strategies of Women in the Informal Labour Sector of Mexico City

The informal activities of women in Mexico City cover a broad range from private tutoring in foreign languages to dishwashing. However, a definite stratification of such tasks exists, related to at least two clearly defined social groups: middle-class women with certain educational and social advantages, and working-class and 'marginal' women with no schooling.

Mexican middle-class women generally consider work outside the home undesirable. According to social norms, their fulfilment, dignity, and respectability lie in home and children. Only a minority, usually university graduates, accept

Table 9.3 Composition of marginal labour force by sex. *Source* Muñoz/de Oliveira/Stern (1972: 258–325)

Occupation	Women (%)	Men (%)
Street peddlers	40.2	59.8
Unskilled workers in services (mainly salaried domestics)	71.8	28.2
Unskilled production workers	21.7	78.3
Unskilled construction workers		100.0
Peasants and agricultural peons	5.8	94.2

salaried work, notably of a professional kind, as a part of a woman's life. Thus it is only women whose husbands do not earn enough who generally engage in part-time activities to earn money. Most can be done at home, or in other women's homes, such as baking, embroidery, sewing or knitting, private tutoring in languages or school subjects, and craftsmanship, such as dolls, boxes, paintings, or leather work. Interestingly, all of these activities are normally done freely for friends and relatives; the only difference between such informal jobs and domestic or family tasks is wages.

Self-employment by middle-class women outside the home usually involves having a small restaurant, or a small shop that sells cakes, china, flowers, knitting, children's clothes, books, and records. A more traditional business is the neighbourhood shop that sells everything from charcoal and candles to flour and sweets. The old shopkeeper who knows all the local gossip is still an institutional figure in some neighbourhoods today. The women who can afford it might establish small song, dance, gymnastics, yoga, or 'personality' academies—a full-time activity for the owner, a part-time activity for women teachers. Significantly, the clientele and the workers in these establishments are almost exclusively women; they constitute an all-female supply and demand labour market.

Lower-class women, in contrast, carry out their activities primarily in the streets or in other women's homes. These activities involve personal services and trade, but most especially domestic service. Domestic service functions as an economic safeguard for migrant and poor urban women because it is always available. However, because it partakes of the traditional female preoccupations with house, children, and kitchen, domestic work is not conceptualized as 'a job'. Since there is no contract, women can be fired with ease and can enter or leave domestic service at various times during their lives. Young women become chambermaids or nannies while they wait to marry; unmarried or separated mothers become domestics in order to support their children. Women between the ages of 35 and 50 can still get work as servants, provided they have no more than one child and no man; only rarely is the husband employed in the same house as gardener, chauffeur, or handyman. It is extremely rare for old women to enter domestic service; the possibility of inefficient work and illness makes employees reluctant to hire them.

Would domestic workers accept an eight-hour work schedule with higher wages if they were offered alternative employment? Given the fact that a female servant's wages are far below the official minimum salary, that she must be available for work day and night with only Sundays off, and that she is often mistreated, one could reasonably assume that she would be eager for another type of job. Indeed, most young women in domestic service today say that they would like to find other types of jobs. However, free room and board for herself and sometimes for her small child often makes domestic work a woman's best option.

Some women live with their husband or children and provide part-time household services, such as washing clothes, windows, and floors, mending clothes, looking after children or the house, and helping out at parties and celebrations. They go to the employer's house, sometimes to seven different houses a week, and return at night to their own homes. Such jobs are sought after and found through friends and relatives. Young lower-class girls can sometimes get a job in a factory or a shop, especially if they have 'good appearance' *(buena presentation).* Or they help in the family enterprise—a restaurant, market stall, or shop.

Middle-aged and elderly women tend to go into petty trade or the sale of food in the streets. Whereas female petty trade usually involves edibles, such as sweets, chewing gum, fruit, and chocolates, in keeping with the traditional image of women as providers of food, men vendors sell clothes, belts, jewellery, and toys. Some women are provided with small carts by the city government; unauthorized street sellers, who move fast to avoid the police, are predominantly rural migrants, many of them Indians. Still other women sell food that they cook in charcoal burners outside their homes, near bus stations, sports grounds, university campuses, or factories. These two types of street vending have proliferated in Mexico City in recent years. However, they tend to offer an unnecessary service and to create their own demand, since a street vendor who sells what she just bought in the market two blocks away is not fulfilling a real demand. It is reasonable to suppose that if they had other alternatives, these women would not engage in such activities.

Data from my own research on a group of underprivileged migrant Indian women in Mexico City suggest that they have few choices open to them. These Mazahua peasants, who started migrating from their villages some 300 km outside of Mexico City in the 1960s, cannot find alternative work. Because most of the men work only intermittently as construction peons, market porters, and in similar low-paid, informal positions, women are obliged to earn additional income through street selling. Widows and divorced or abandoned women who have migrated to the city with their children are also street vendors. Their blatant poverty has aroused public attention and led to a repression of their activities by city police. City officials have wondered why these women have not conveniently disappeared into domestic service. Indeed, domestic service is used as a shield against protest over women's high unemployment rate and over the government's unwillingness to do anything about it. A simple analysis shows the limited range of occupational choices open to lower-class women in Mexico City. Most of these migrant women have children and/or husbands and are thus disqualified from residential domestic service. Even if they were willing to be separated from their families, their wages would be lower than those paid to urban women, and they would probably be ill-treated by their employers. Irregular domestic service is also unavailable to them because they lack the necessary skills and the social contacts to find such jobs. As dishwashers and kitchen help in restaurants, they know that they would be exploited, sometimes working up to fifteen hours a day for extremely low pay. Young urban girls are usually preferred as waitresses. Finally, these migrant women lack the knowledge to sell in established markets; this requires an appropriate license for a stall, an ability to cope with market administrators and inspectors, and established business contracts to obtain their merchandise at wholesale prices. Outright begging is subject to police harassment, and Mazahua women consider it degrading. Thus, they prefer to sell fruit or sweets in the streets, an activity which they already know from their peasant villages. The network of former Mazahua migrants in the wholesale fruit and vegetable section of the

Merced, the main city market, gives Mazahua women access to wholesale prices. In addition, street selling can be taken up and left at any time, either in order to return to the village or to stay at home when a child or husband is ill. Another advantage is that small children remain with their mother while she sells, at times that are convenient to her and in places where she can talk and joke with her friends and relatives. Even more important, income from sales in 'good spots' is higher than her husband's unstable earnings as a construction worker or porter. Thus some husbands have stopped working altogether and only help their wives out by carrying the crates of fruit to the house, or by hiding the bulk of it in case the police arrest their wives. Others simply wait for their wives to bring home some money with which to get drunk. As this brief description suggests, given their occupational options, street selling provides the greatest advantages to these migrant women. Guided by decisions that are very rational, they make the most out of their hopeless, underprivileged situation.

Finally, another informal occupation open to women of all classes is prostitution. Data on prostitution are virtually non-existent. According to one unpublished psychiatric study conducted in Mexico City, prostitutes are either extremely neurotic or mentally retarded, an explanation given in medical circles throughout the world. Although a serious attempt at a sociological survey was stopped by gangsterial opposition, bits of information suggest that young girls from middle-class families occasionally go into prostitution before marrying because they can make more money in one evening than in two weeks of secretarial work. For some, prostitution is a reaction against the tediousness and repression of their family life. The lack of reliable studies means that we can only speculate about the reasons why lower-class women become prostitutes. Judging from occasional data, it would seem that prostitution is a major informal source of income for women, but it is never taken into account in discussions of female economic participation or survival.

9.6 Conclusions

Informal activities in developing capitalist economies must be understood within the total pattern of employment in a given country. Whether temporarily or permanently, informal activities are usually taken up when formal jobs are unavailable. For women especially, the dividing line between formal and informal jobs is, as we have seen, very tenuous. On the one hand, formal employment implies a long-term, full-time contractual job, but this does not apply to paid domestic work, petty trade, and small craft production, all of which have large numbers of female workers. These jobs are considered informal if they are carried out intermittently on an irregular basis. On the other hand, the boundary between women's wage labour in the informal sector and unpaid household and community tasks is even more tenuous. A woman not needing an income will do exactly the same tasks—baking, sewing, embroidery, crafts, and tutoring—without pay that other women who need an income do for a wage. Indeed, at the formal end of the spectrum, women servants who earn a salary are considered economically active and are included in the GNP; midway, women do the same tasks on an informal basis—part time and intermittently—and yet are paid for them without being considered economically active and without being included in the GNP; at the other end of the spectrum, housewives do full-time domestic work with no payment and are considered outside the economic system. All of these women obviously are doing exactly the same type of work. Thus there is a very real need to reconceptualise women's work.

The analysis of female work in Mexico City has suggested that when women are in economic need, they press the system for payment of their domestic services—but not those done for their own families. Instead, women turn domestic duties into economic activities by offering such services to other women. This represents an all-female supply and demand market. At the same time, our analysis indicates that as long as women can enter paid domestic service, female unemployment will not be officially or publicly acknowledged. Government officials can claim that there is no female unemployment problem since lower-class women can always become servants.

As the data on Mexico City suggest, women withdraw from the labour force by the age of thirty to raise a family, but later on are unable to re-enter the labour force, primarily because of the unavailability of jobs. Since women with schooling have higher unemployment figures than their male counterparts, education, contrary to what is frequently argued, does not represent the determining factor in women's unemployment. Age is an important factor; expanded job opportunities in Mexico have gone to young women between the ages of fifteen and thirty. After thirty, the census indicates, women have more difficulty finding jobs than men. In such a situation, the notion that women should remain at home will tend to be reinforced, even by women themselves. Data showed an inverse correlation between women's unemployment and their participation in domestic tasks. Importantly, this means that female unemployment figures are far from accurate, since many women stay on sharing household duties rather than consider themselves unemployed when they are unable to find work.

When women who need to work cannot find jobs, they compensate for their unemployment by taking up informal activities. Middle-class women take up jobs they can carry out in their own homes or in those of other women. The largest number of women in Mexico City, though, the lower-class women, usually have to engage in informal activities even if their husband has a steady income. And of course, elderly, divorced, widowed, or single women depend entirely on informal activities. The ever-increasing number of women who sell fruits and sweets or snacks in the streets do not fill a demand within the urban economy, but create their own demand out of desperation, since they have no other economic alternative. Immigrant women in Mexico City have the most restricted range of options. The Mazahua women street sellers merely choose the activity that gives them the most advantages within their set of options. Indeed, if we were to ask them why they chose to work at such a low-income, harassed activity, most probably they would answer, "Well, one must carry on one's little fight, mustn't one?" [Pues tiene uno que hacer su luchita, ¿no?].

References

Boserup, Ester, 1973: "Employment of Women in Developing Countries", Paper presented at the International Population Conference, Liege.

Cardoso, Fernando H; Faletto, Enzo, 1973: *Dependencia y desarrollo en América Latina* (Mexico City: Siglo XXI).

Chaney, Elsa; Schmink, Marianne, 1975: "Las mujeres y la modernización: Acceso a la tecnología", in: Elu de Leñero, María del Carmen (Ed.): *La Mujer en América Latina*, vol. I (Mexico City: Sepsetentas): 24–54.

de Oliveira, Orlandina, 1976: *Absorción de mano de obra a la estructura ocupacional de la Ciudad de México* (Mexico City: El Colegio de México).

Frank, Andre Gunder, 1972: *El Desarrollo del subdesarrollo* (Mexico City: Escuela Nacional de Antropologia e Historia).

González Salazar, Gloria, 1975: "La participación de la mujer en la actividad laboral en México", in: Elu de Leñero, María del Carmen (ed.): *La Mujer en América Latina*, vol. I (Mexico City: Sepsetentas): 105–135.

Hobsbawm, Eric, 1975: Age of Capital 1848–1875 (New York: Charles Scribner's Sons).

Hobsbawm, Eric, 1982: *Marxism in Marx's Day* (Bloomington-Indianapolis, IN: Indiana University Press).

ILO (International Labour Organization), 1975: *Womanpower* (Geneva: International Labor Organization).

Muñoz, Humberto; de Oliveira, Orlandina; Stern, Claudio, 1972: "Migración y marginalidad ocupacional en la Ciudad de México", in: Martínez Ríos, Jorge (Ed.): *El perfil de México en 1980* (Mexico City: Siglo XXI): 325–358.

Myrdal, Gunnar, 1968: *Asian Drama* (New York: Twentieth Century Fund).

Nun, José, 1969: "Sobrepoblación relativa, ejercito industrial de reserva y masa marginal", in: *Revista Latinoamericana de Sociología*, 69,2: 178–237.

Quijano, Anibal, 1970: "Redefinición de la Dependencia y Proceso de Marginalización en América Latina", Comisión Económica para América Latina (Mimeo).

Rostow, Walter Whitman, 1960: *The Stages of Economic Growth: A Non-Communist Manifesto* (Cambridge: Cambridge University Press).

Ruíz Harrell, Rafael, 1975: "Aspectos demográficos, educativos y laborales de la mujer en México, 1960–1970", Paper presented at the International Women's Year Conference, Mexico City.

Saffioti, Heleith, 1975: "Relaciones de Sexo y Clases Sociales", in: Elu de Leñero, María del Carmen (ed.): *La Mujer en América Latina*, vol. II (Mexico City: Sepsetentas): 35–59.

Schmink, Marianne, 1974: "Dependent Development and the Division of Labor by Sex", Paper presented at the Fifth National Meeting of the Latin American Studies Association, San Francisco.

Singer, Paul, 1975: *Economía política de la urbanización* (Mexico City: Siglo XXI).

Stavenhagen, Rodolfo, 1972: *Sociología y subdesarrollo* (Mexico City: Nuestro Tiempo).

Suárez Contreras, Enrique, 1973: "Migración interna y oportunidades de empleo en la Cd. de México", in: Martínez Ríos, Jorge (Ed.): *El perfil de México en 1980* (Mexico City: Siglo XXI): 359–418.

Tienda, Marta, 1974: "Regional Differentiation and the Sector Transformation of the Female Labor Force: Mexico, 1970" (Ph.D. dissertation, University of Texas).

Youssef, Nadia, 1971: "Social Structure and the Female Labor Force: The Case of Women Workers in Muslim Middle Eastern Countries", in: *Demography*, 8,4 (November): 427–439.

Chapter 10
Cultural Change and Ethnicity on Rural Mexico

An unease with the term 'cultural change' seems to be characteristic of research in rural areas in Latin America in the last few years. It stems, in my view, from two sources: first, from the confusion arising between the holistic term of culture as used in classical anthropology, and the reductionist version of it prevalent in other social disciplines and in policy-oriented studies, and, secondly, from the belief that the concept of cultural change cannot be dislodged from the North American culturalistic framework, and thus cannot be applied in studies taking a Marxist or dependency theory framework.

Because these theoretical difficulties have not been made explicit, very different approaches have been taken in different studies. In some, cultural change has been subsumed into a more general process of 'modernization'; in others, it has narrowed in focus into ethnic change; and in others still, it has been left aside, since the implicit assumption is made that the ideological superstructure (that is, culture in a restricted sense) is nothing more than a mechanical reflection of the economic infrastructure, and thus efforts should be directed at explaining the latter.

This paper attempts to give a comprehensive view of cultural change in the Mazahua region in central Mexico, since the beginning of this century, by focusing on the association between mainstream economic processes and shifts in ethnicity, in literacy, and in the perception of rural culture in the region. In so doing, it is hoped that both narrowing down the concept and separating it from the economic and power structure will be accented.

Studies of cultural change in Latin America indeed began with a culturalistic approach. The Malinowskian scheme of culture change was a major influence and was applied in Mexico (Foster 1960). Later, cultural change came to be known as 'acculturation', a term meant to bridge the gap between otherwise static ethnographic accounts of cultures. But 'acculturation' could not occur unless it was assumed that

The chapter was originally published in Preston, David A. (Ed.), 1980: *Environment, Society and Rural Change in Latin America* (Chichester: John Wiley & Sons): 123–134. The permission to republish this text was granted on 4 July 2013, by Verity Butler, Permissions Co-ordinator, Wiley, Chichester, UK.

L. Arizpe, *Lourdes Arizpe*, SpringerBriefs on Pioneers in Science and Practice 10, 115
DOI: 10.1007/978-3-319-01896-6_10, © The Author(s) 2014

cultures with discrete boundaries existed which were becoming other cultures. Thus, indian cultures were assumed to exist, as direct inheritors of a pre-Hispanic past. Under the weight of such assumptions, however, it became increasingly clear that the social groups and processes that could explain the direction, intensity, and selectivity of social change were lost (for criticism of this approach see Stavenhagen 1974; Pitt-Rivers 1973; Van den Berghe 1974; Ribeiro 1968; Bonfil 1972).

In the attempt to overcome the difficulties posed by this culturalistic approach, modernization theory has attempted to provide a comprehensive scheme of change from 'traditional' societies to 'modern' societies, which could be applied to all rural areas undergoing changes around the world (Lerner 1958). This dual scheme for societies in the Third World, though, has been severely criticized (among others, by Singer 1975; Ribeiro 1975). The main difficulty, it seems to me, is that modernization theory does not add any analytical insights. It narrates a process and divides it up into diverse stages, again and again. But even if it is divided into a thousand stages, it will still be the same scheme, one which, bowing to Redfield's heritage, fundamentally assumes economic change. Redfield intended to provide a blueprint for *social* change with his folk-urban continuum, but instead he provided one for *cultural* change in a restricted sense of the term. It was not the *technical* order that concerned him: he assumed that the peasant economy would by and large evolve into an 'urban' type of economy. What interested him was to explain the *moral* order. In the same way, in modernization theory it is assumed that traditional peasant economies are being integrated into national and international market systems. Of course, there is this constant transformation going on, but the vital issue to explain is, surely, why changes occur *differentially* between nations, regions, and villages, and along class lines, ethnic lines, and sex lines. Thus, my proposition is that, once the major economic trends of capitalism in peripheral countries are known, attention should turn from trying to explain the uniformities of this process to explaining the disparities within it. This, it seems to me, is especially important in the relationship between rural and urban areas at the local level.

Lourdes Arizpe and 'Vaquero: at the Festivity of 'Toro Petate' in Cuajinicuilapa, coast of Acapulco, Guerrero, in Mexico in 2009. *Source* Author's personal photo collection

Unequal development at all levels, from the local to the international, is a consequence of the dynamics inherent in capitalism. At the local level in rural areas, this theoretical approach has tended to postulate a mechanical relationship between core and periphery which has already been contested, correctly in my view (Roberts 1974; Kemper 1970), but no alternative analysis of the relationship of culture to mainstream economic processes has arisen. This relationship is a highly complex one and cannot, at this time, be understood in overall terms. The first steps in this direction will have to be to point at areas of contradiction and of reciprocal influence.

That cultural change in rural Latin America has closely responded to events on the national scene is becoming more and more evident in historical studies. For example, in Mexico, already in the eighteenth century, the apparently isolated rural areas responded to the widespread economic recession by the formation of closed, corporate indian communities that consolidated a 'traditional' way of life (Wolf 1959). Seen from this perspective, traditionalism, as several authors have noted, becomes a symptom, rather than a cause. However, at a theoretical level, it is necessary to postulate that at certain historical periods cultural phenomena can, in fact, overdetermine economic processes.

In the Mazahua region in the last 50 years, cultural change has gone ahead in stops and starts, differentially for ethnic groups and for social classes, with geographical disparities between villages. Although one same trend is evident at the regional level, this paper analyses the disparities and factors underlying it.

In this paper, the terms 'culture' and 'culture change' will be used throughout in a restricted sense, referring primarily to modes of thought, language, art, customs, and rules of social behaviour.

10.1 The Mazahua Region

The Mazahua region encompasses eleven municipalities in a high arid plateau 300 km north-west of Mexico City with a total population of 200,000 of whom roughly half speak the Indo-American Mazahua language. It is an agricultural region of *ejido* lands (lands owned by the State allotted to the peasants), with an acute fragmentation of land and low productivity that is leading to a rapid proletarianization of the peasants. As a result, both the *mestizo* and the Mazahua households are involved in seasonal and permanent out-migration, mainly to Mexico City and the Mexico-United States border. Migration and the spread of mass media and literacy have brought a strong cultural impact to the region, building up pressure for the changing of values and cultural habits. Cultural change in the last 50 years, though, has occurred differentially along ethnic and class lines.

10.2 Ethnic Groups in the Region

At the turn of the century, the social structure of the region, whose pivots were the export-oriented *haciendas,* clearly allocated roles in production along ethnic lines. It may be even more correct to say, according to what informants stated, that ethnic identity was allocated by position in the production system. The owners of the *haciendas,* informants say, were all of Spanish descent. In fact, only one family, De la Fuente y Parres, owners of La Providencia, the largest *hacienda,* was of such descent; other *hacendado*'s families were assigned this putative descent because it was appropriate to their identity as the landowning class.

The foremen and administrators were *mestizos* or impoverished whites. Only one informant mentioned this distinction, all the others considered them as *mestizos*. Thus, no difference was made between these two groups, either in social or in cultural terms. The boundary drawn between them and the labourers and peasants, however, was strikingly clear—the latter were indians. That this distinction was in no way racial is shown by the fact that children, such as the illegitimate offspring of the *hacendado*'s sons, and the *mestizo's* legitimate or illegitimate offspring—frequently resulting from the rape of Mazahua women—who were brought up in the *hacienda* household became *mestizos*. Those, of the same parental origin who were raised among the Mazahua became indians.

The Mazahua, in turn, called and still call the other group *mestizo* or *cruzado*—'mixed'—and call themselves *mazahueros.* They refer to their nearest geographical and linguistic neighbours by the name of their language: the Otomí—to the east and north—and the Matlatzinca—to the south. They consider them very different from their own group, and only rarely will they call them *hermano indio.* The *mestizo*, in contrast, make no distinction among these groups, and call them all indians. This coincides with reports from other places such as Guatemala (Van der Berghe 1974).

In *hacienda* times, there was no distinction between the *peones acasillados,* the Mazahua labourers who lived on the *hacienda* premises, and those who lived in peasant households in the villages. The former were a specialized workforce who laboured in the workshops where the root of the *zacatón*— a grass—was processed for export to Europe and the United States.

Interviewing the 'Malinche' in the Dance of the Conquest performed in Tlacoachistlahuaca, Oaxaca, Mexico in 2008. *Source* The author's personal photo collection

The Revolution of 1910 came to the region entirely from the outside. The *hacendados* fled abroad, the *mestizos* entered the fray, and the Mazahua fled to the nearby hills. By 1925, the only major apparent results of the fighting in the region had been deaths, famine, and epidemics. The *haciendas* had missed a few steps but still continued production. It was not until the Federal Government decided to distribute the land that the *mestizos*, organized into *agrarista* groups, invaded the *hacienda* lands.

By the end of the 1920s and in the early 1930s, the Government had allocated *ejido* lands to almost all the villages in the region. Significantly, having obtained the lands, the *mestizo* and the Mazahua populations agglutinated into new settlements along ethnic lines. At this point, it seems to me, ethnic loyalty overrode other considerations. If the creation of the ethnic identity of the Indian was fostered and perpetuated by assigning Indo-American peoples a distinct economic role in the colonial economy, and later on in the *haciendas,* the fact that Mazahua cohesion persisted after these pressures subsided shows that their ethnic identity was by no means only an economic specialization. Interestingly, the data show that, from that point in time on, a dialectical relationship was established between ethnic filiation and economic pressures.

The Mazahua villages got their land and immediately retracted into corporate, isolationist units based on a household economy of the same type that Wolf (1959) describes in rural areas in eighteenth-century Mexico. It is significant that, in almost all Mazahua villages, this coincided with the rise of *caciques* who deliberately and ostentatiously proclaimed their Mazahua identity. In the *mestizo* villages, in contrast, the *caciques* doubled as local and regional representatives of the ruling party and new entrepreneurs in trade and in commercial agriculture. Thus, the economy of these villages rapidly underwent the changes attendant on the spread of capitalism into this rural region.

In the 1930s, both ethnic groups received equal amounts of land, and capital goods were distributed more or less equally between them. By the 1970s, the Mazahua were clearly the poorer of the two. Elsewhere (Arizpe 1978), a detailed analysis is presented of the social, economic, political, and cultural mechanisms that help explain their progressive impoverishment. All peasant households in the region have felt the impact of economic forces that destroy the peasant economy, but for the Mazahua the consequences have been more acute. Women, especially, have lost sources of income in traditional handicrafts, cottage industries, petty trade, and traditional occupations such as midwife and *curandera* (healer). Men also lost occupations such as plough carver, musician, bone setter, and others, as well as wage labour in the mines and in the *hacienda* fields. Sources of cash income for the households declined while their needs for cash soared. The sons and daughters of *mestizo* families, because they benefited from schooling and from social priority in entering employment, have taken on the new occupations generated by capitalist development such as seamstress, tailor, teacher, and electrician, and have taken up new economic opportunities to set up small business such as shops, restaurants, and taxi and bus services. Since so many of them left agriculture, their brothers—women were legally barred from inheriting *ejido* lands until 1975—were able to inherit sizable amounts of land. In contrast, the Mazahua younger generations are locked into the agricultural sector and have had to inherit ever more fragmented plots of land. This has led the domestic units to more acute decapitalization and proletarianization.

At the national level, important cultural events exerted a strong influence in the region. At the end of the 1920s, Jose Vasconcelos crystallized a new concept of the Mexican nationality: the *raza cósmica,* a cosmic race that would successfully blend the Indo-American and the Spanish heritages. Since the indo- American groups were allocated *ejido* lands with no ethnic distinction being made, the new Government expected 'acculturation', a term just then beginning to be used in academic circles, to occur naturally and automatically. As the data from the Mazahua region show, in spite of this open invitation, the indian groups remained indian. In the 1940s, as it became increasingly apparent that these indo- American communities not only were not integrating into the economic dynamic of the new society, but continued to be as exploited and repressed, at the local level, as ever, the Instituto Nacional Indigenista was established. The avowed purpose of this institution was to integrate the communities into the national economic and social system. This policy of opening indo-American communities in order to achieve development by, among other things, enlarging the internal market, couched itself in cultural terms and, as will be seen further on, was responsible for building up pressure to make the indo-Americans reject their indian identity.

Although there was no direct action of the Instituto Indigenista in the Mazahua region, the policy of *indigenismo* (indianism) made itself strongly felt, as can be seen in this excerpt from an interview with a Mazahua man: asked what he thought about the fact that the Mazahua language and customs were disappearing he answered:

> But, lady, isn't that what the government wants? Because we speak here but only here with those who live in the village. We should rather speak as they do elsewhere, as the nation does. So we shall make a fatherland.

Further on in the interview, he stated explicit reasons for wishing his children to learn Spanish:

> This would allow them to do better; oh yes, he (his child) can go to ask for a job and the boss asks him—'what knowledge do you have?'—and if he has no knowledge they don't give him the job. On the other hand, if he can express himself, they say to him—'come here and get yourself some pennies'.

Ethnic filiation also influenced the strategies taken by the peasant households faced with progressive proletarianization. Migration became the most important source of income for both groups, with offspring being increasingly sent off, by order of birth, to wage labour. *Mestizo* young men migrated at first towards the Mexico-United States border to work as agricultural labourers in the United States. After the *bracero* programme (the official treaty between the two Governments) was terminated in 1964, they went to Mexico City where they were generally able to get jobs in factories, in offices, or in the services sector as bus or taxi drivers, agents, or salesmen. *Mestizo* young women also went to Mexico City and usually got jobs as shop assistants or in offices. Many of them, both women and men, settled permanently in Mexico City.

In contrast, during the 1950s, the Mazahua men began migrating in search of agricultural work to various regions and to Xochimilco, an agricultural suburb of Mexico City. They rarely settled in these places, since their primary aim in migrating was to get additional income for their parents' household. In the 1960s, they began to migrate increasingly to Mexico City to work as market porters, masons, and as assistants in La Merced market stalls and warehouses. The Mazahua women, almost without exception, worked as domestics and rarely had occupational mobility. They usually left the job to get married, usually to someone from the village.

Again, the fact that the Mazahua migrants normally returned to their village increased population pressure on the land within that group. Some Mazahua villages experienced permanent out-migration. In the 1950s, Mazahua migrants were able to escalate into formal permanent employment because of the overall expansion of the manufacturing and services sector in Mexico City. In the 1960s, their mobility slowed down as the expansion waned. Skill and documentary requirements increased for factory work, and the Mazahua then began to stay within their ethnic community in the city. That is, the dialectic that made 'passing' into the urban culture advantageous came to a standstill.

In terms of ethnicity, two strategies emerged to cope with unemployment and occupational immobility. First, the community specialized in a trade activity. Mazahua women intensified their street selling and retained their Mazahua identity because it became useful in competing with other city vendors. Second, many of the migrants joined a new religious brotherhood, the *Concheros,* in what seems to me to be an effort at recreating ethnic solidarity—and through it, financial and social reciprocity—in an otherwise fluid and unstructured cultural context.

10.3 A New Urban Ethnic Identity

The *Concheros* originally came together to dance purportedly Aztec dances at church celebrations in Mexico City. The group, named after the *conchas* or shells they use to accompany their music, became a brotherhood in which initiated members address each other as *hermano* and *hermana* and engage in highly ritualized and hierarchized social relationships.

With an ease of creativity that other social classes in Mexico society certainly deny themselves, these people have blended together Aztec symbols and designs, ritual elements of the *mayordomías* and *cargo* systems of some of their rural villages, Catholic religious liturgy, and the oratory and pomp of political ceremony, to create a new, eloquent image of their collective identity. Their costumes are sometimes luxurious and beautiful, their conviction and en-thusiasm contagious. Their new identity is probably more genuine than that of the urban middle classes who derive theirs passively from the mass media.

Most of their members are of rural origin, with a great number of Otomi and Mazahua among them, and many of them have lived in the city for 15 years or longer. The men work at low-income jobs such as street vendors, doormen, lottery tickets salesmen, night watchmen, and the like. The women do domestic labour as well as part-time informal activities, street selling, sewing, and washing clothes. For almost all of them, the city has not kept its promise of providing a permanent, well paid job. Caught between the rural cultures they left behind and the urban culture dominated by the middle class, and which is out of their reach, it seems to me that they have to revert to purported indian and Aztec symbols to form an immediate reference group and a new identity. In the older *barrios* or *vecindades* of migrants of Mexico City, such subcultures have always been created. For example, the subculture of the *peladito,* so well portrayed by Cantinflas, the Mexican comic actor, or the subculture of 'Tepito'. In the case of the *Concheros*, the fluid residential mobility and the new shanty towns have not been able to provide a *barrio* basis for a new subculture. Thus, they have formed their own by reactivating rural ties of before, and the Aztec culture.

Although urban-based, *Conchero* groups frequently go to dance in the villages of origin of their members. For example, they danced at the patron saint's day fiesta in Santiago Toxi, a Mazahua village.

In the view of the urban middle class, they belong to an indian, 'backward' culture. In fact, it is a *new* indian subculture created for specific purposes of cohesiveness and solidarity in a social space that has nothing to offer but a cultural vacuum.

Interestingly, many *Concheros* are literate and have had some schooling. In fact, it is from school textbooks that they have extracted elements of the Aztec culture to use as their new symbols.

10.4 Schools in the Region

The role of the schools in cultural change in the region can be illustrated by data from two Mazahua villages. Primary schools in Santiago Toxi and San Francisco Dotejiare opened in 1945. In both, Mazahua children stayed away from the schools in Toxi until the late 1950s and in Dotejiare until the early 1960s. The reasons cited for this were that schoolteachers and *mestizo* children discriminated against, and often mistreated, Mazahua children. Also, informants said, the children were needed to help in the homestead. Boys and girls had to tend the flock of sheep, girls had to help with the domestic work. Female children, in particular, were not allowed to go to school, not only for the reason just cited but because it was feared that they would be raped, a fear totally based on real possibilities. Even today, young girls are taken out of school as soon as they begin to menstruate. And also, it was thought that women did not need an education.

Among the *mestizo* families, both male and female children attended school. The rape of *mestizo* girls is less frequent, because of the brutal retaliation that men in their families would exert. *Mestizo* parents were always very conscious of the advantages inherent in acquiring skills, and sometimes sacrificed their need for additional labour in the home so that their children could attend school.

These two attitudes could be interpreted as stemming from different cultural traditions. But the differential access of the two ethnic groups to employment and to economic opportunities show that both were rational decisions within the context of each group's sphere of action. Mazahua children could not enter wage employment in the services of offices in the villages or towns. School curricula were strikingly urban in content so that, in fact, they could not apply what they learned at school to their agricultural activities. The *mestizos*, in contrast, could apply this knowledge, since they would be dealing with national institutions and with jobs that required skills learned in the schools. Moreover, for Mazahua children, attending school usually meant a change in ethnic filiation, since teachers forbade them to speak their language and to wear Mazahua garments.

Since they were to live in the villages, such an estrangement from the Mazahua community was detrimental to their chances for survival and mobility.

This is supported by data from Toxi showing that Mazahua children began attending school massively during the second half of the 1950s. It was in this village that population pressure on the land was greatest, so that migration to Mexico City began earlier than in other Mazahua villages. Also, families had little use for all of the labour of children, having lost their herds and subdivided their lands. Success in wage labour in Mexico City, unlike the local situation, depended on a good knowledge of Spanish and basic skills in reading and arithmetic.

For example, a young woman or man who could keep accounts *(hacer cuentas)*, got a much higher wage than other helpers who only carried merchandise or waited on customers. The migrants that returned to their villages of origin took with them this pattern of thinking, relating skills to economic mobility.

At the same time, nearby, a large industrial complex was established in the 1960s which gave young people in the region access to factory employment. Management in the factory very early on realized they had to carry out a deliberate campaign in the region fostering cultural change to 'civilize' the people in the localities (Arizpe 1978). This means to turn peasants into factory workers, or to be more precise, to have young people adopt appropriate attitudes that would provide the rationale for them to accept the conditions of factory work. The campaign to foster cultural change was successful and the firm has brought in more than twenty million pesos annually to the area in wages. As a result, the whole life style of the area has changed. Among other things, school enrolment jumped in Toxi at the beginning of the 1970s, to such an extent that parents are now sending their children to the school in a nearby town, Ixtlahuaca, because the teachers are said to be better there.

In contrast, in Dotejiare, the only wage labour available is in agriculture. Opportunities in commerce or transport are monopolized by the Mazahua elite and a few *mestizo* families. Migration among the peasant households follows a strictly family and kin type of strategy. Land is still available, so that the younger generation can return to agricultural activities. As a result, school attendance has to be induced by the school principal and the *cacique*. Even families where children's labour is not needed send their children to school reluctantly. Dotejiare migrants in Mexico City have tended to keep within the protective 'shell' of Mazahua identity. Thus, ethnic ties are still more important than skills in ensuring economic survival of mobility.

10.5 Government Policy and Rural Culture

The Mexican Government has not formulated a policy of cultural change since Vasconcelos and the 1930s, when it generally advocated the spread of literacy, a strong nationalism in the arts, and indianism. Subsequent regimes have taken a *laissez faire* attitude regarding culture, while attention has been focused almost exclusively on economic growth. At the local level, the result of this lack of policy can be seen in the drop of the quality of schooling and a feeling of aimlessness in Government projects directed at culture, such as the *Misiones Culturales* (the social workers sent to the villages in the 1940s and 1950s to foster cultural change), and in a decline of cultural expression in the region as a whole. Documentary evidence indicates that at the beginning of the century, the larger towns in the region had some cultural life: local people wrote on its history, its landscapes, and the lives of those around them.

The political centralization of the new post-Revolutionary Government in the 1920s reflected itself in a marked cultural centralization in Mexico City. Interestingly, the novels, films, and mural art of the 1920s and 1930s still had their roots in rural society: they dealt with the *hacienda,* the Revolution, and the *charro,* the national figure. Songs and singers of rural origin were promoted, and films acted by Dolores del Río, Maria Félix, Jorge Negrete, and others, often under the masterful direction of Emilio 'Indio' Fernández, vividly portrayed the passion and violence of the countryside. In the 1950s, the revolt against an excessive nationalism in the arts and the rise of the urban middle classes shifted the pivot of Mexican culture to the city. Whereas before, in the 1940s, both the rural and the urban were reflected in films and novels, in the 1950s and 1960s, the urban has predominated to the almost total exclusion of the rural.

At the local level, this can be clearly traced. Up until the 1950s, the radio constituted the main cultural linkage to Mexico City. According to those who remember, programmes were good: the *locutores*—radio broadcasters—read poetry and commented on events; the *radionovelas*—serialized melodramas—were very popular. But songs and singers of an exclusive Mexican heritage formed the core of cultural transmission on the radio.

By the end of the 1950s, the quality of radio programmes declined as attention began to turn towards television in the city. Slowly, programmes on the radio became mostly music, much of it American, and news broadcasts became sensationalist. The favourite programme on the radio in the Mazahua region in the 1970s was an 8 a.m. news programme containing information about the latest murders in the vicinity.

As literacy, or rather, semiliteracy increased, the region was deluged by cheap comics and *fotonovelas* (melodramatic stories in the cartoon format incorporating a variety of subjects ranging from adventure to romance and passion).

Television made its appearance on a larger scale at the end of the 1960s. Few families have one, but they charge fifty cents to other people who want to watch it, and these are mostly children. Television programmes are in a majority serialized melodramas or *telenovelas,* and dubbed American- televized series. Viewing time is very heavily loaded with commercials showing a 'modern' way of life pointing towards consumerism. In effect, what has happened is that the mass media are now in total control of cultural change in the region. But they beam out an almost exclusively urban culture. The effect of this 'modernization' has been a sense of alienation in the region: young people, particularly, despise the rural way of life and everything having to do with agriculture. As a result, cultural change is not shaping itself to fill the needs and configuration of a society that derives its livelihood from agriculture, but those of the urban one in Mexico City. The rural culture has been robbed of its dignity and its future: there is no place for it in the charter of development for the Mexican society.

10.6 Conclusions

Ethnic filiation has directly affected peasants in the Mazahua region by channelling forces that destroy the peasant economy, on the one hand, and that provide wage labour and allow capital accumulation, differentially among ethnic groups. The Mazahua indian group has had to bear the impact of the former, yet it has been barred from access to the latter. The volume and speed of 'passing' from the Mazahua ethnic group to the *mestizo* group was seen to be directly affected by economic opportunities and by Government policy favouring this process. In general terms, it was shown that the existence of an ethnic boundary channels the process of proletarianization so that the class line will again coincide with the ethnic line.

Analysis clearly indicated a dialectical relationship between economic mobility and ethnic filiation, both in the rural area, as well as in the city. In the latter, migrants depend on kin groups and the ethnic community to get jobs, housing, and financial help. If overall expansion of employment allows them mobility, they tend to leave the ethnic community and to change their cultural habits. If occupational mobility is not possible, then they tend to stay within the group and its economic specialization. However, reciprocally, those who, from the be-ginning, never venture outside the work and social area of the ethnic group have less possibilities of achieving occupational mobility. The availability of employment is, however, the necessary condition for this to be a relative disadvantage.

Mazahua migrants unable to achieve economic mobility in Mexico City have taken two strategies: either they retain their ethnic identity, enabling them to participate in their economic specialization and to receive collective support in crises, or, if their Mazahua ethnic ties are already too tenuous, they enter a newly created ethnic grouping vaguely rooted in the past, the *Concheros.*

Schools in the Mazahua region are not instrumental in an absolute sense in bringing about literacy and 'modern' attitudes, that is, behaviour appropriate to industrial work and consumerism. School attendance implies a rejection of the Mazahua identity, and the latter will only be abandoned if economic survival and mobility do not depend on belonging to the ethnic community. Only where opportunities for widespread wage employment exist, as in Santiago Toxi, or where constant migration links the village to the pattern of industrial employment, do parents become interested in increasing the level of schooling of their children.

The lack of a coherent, explicit cultural policy emanating from the Mexican Government after the decade of the 1940s has shifted control of cultural change in the country to the mass media, as shown by data from the region. Also, the rural content of Mexican culture portrayed in the cinema, the literature, and the arts until the 1950s has almost entirely disappeared. Mexican culture has become predominantly urban, middle class, and outward looking. At the local level, this has shown itself in the changeover from good-quality radio programmes to poor- quality comics, radio, and television programmes. In all of them, rural life is portrayed as backward and degrading, with no future of its own. As a result, an alienated culture is spreading in the region, more dependent than ever before on outside initiatives.

References

Arizpe, Lourdes, 1978: El Reto del Pluralismo Cultural (Mexico City: Instituto Nacional Indigenista).

Bonfil Batalla, Guillermo, 1972: "El concepto de indio en América: una categoría de la situación colonial", in: Anales de Antropología, 9: 35–45.

Foster, George M., 1960: Culture and Conquest: America's Spanish Heritage (New York: Wenner-Gren Foundation for Anthropological Research).

Kemper, Robert V., 1970: "El estudio antropológico de la migración a las ciudades en América Latina", in: América Indígena, 30,3: 609–633.

Lerner, Daniel, 1958: The Passing of Traditional Society: Modernizing the Middle East (Glencoe, NY: Free Press).

Pitt-Rivers, Julian, 1973: "Race in Latin America: the Concept of 'Raza'", in: Archives Européennes de Sociologie, 14: 3–31.

Ribeiro, Darcy, 1968: The Civilizational Process (Washington: Smithsonian Institute Press).

Ribeiro, Darcy, 1975: Las Américas y la Civilización: Proceso de Formación y Causas del Desarrollo Desigual de los Pueblos Americanos (Mexico City: Nuestro Tiempo).

Roberts, Bryan R., 1974: "The interrelationships of city and provinces in Peru and Guatemala", in: Cornelius, Wayne A.; Trueblood, Felicity M. (Eds.): Latin American Urban Research, vol. 4 (Beverly Hills: Sage): 207–236.

Singer, Paul, 1975: La Economía Política de la Urbanización (Mexico City: Siglo XXI).

Stavenhagen, Rodolfo, 1974: "La Sociedad Plural en América Latina", in: Diálogos, 55 (Jan–Feb): 5–10.

Van den Berghe, Pierre, 1974: "Ethnic membership and cultural change in Guatemala", in: Heath, Dwight B. (Ed.): Contemporary Cultures and Societies of Latin America (New York: Random House): 316–327.

Wolf, Eric Robert, 1959: Sons of the Shaking Earth (Chicago: University of Chicago Press).

Chapter 11
The Challenge of Cultural Pluralism: The First National Congress of Indigenous Peoples 1975

As anthropologists, we are concerned not only with studying the indigenous peoples of Mexico, but also in helping improve their livelihoods and standards of living. My fieldwork notes, then, must also become records of the rise of a new consciousness about peoples' lives in Mexico.[1]

The First National Congress of Indigenous Peoples was held in October 1975 on the shores of Lake Pátzcuaro, an appropriate site, since it was there that Don Vasco de Quiroga made the first attempt to organize the Tarascan communities under a new political and economic scheme. I should begin with a point of clarification: indigenous peoples are Mexican *campesinos* (peasants) who speak native American languages and preserve, to a greater or lesser degree, distinct customs and institutions. These are not all 'prehispanic' since anthropology has just shown that a large part of their cultural patterns were developed after the Conquest. It is a mistake to believe that after this event, indigenous cultures were frozen, to become archeological monuments remains that are neglected. On the contrary, cultures do not stagnate: they continue to recreate ancient elements but they also incorporate new ones, integrating both of them into a spiral of constant development. That every culture has historic validity as well as a vitality of its own is a scientific fact. Instead, it is a political fact that given economic and political interests, some groups may want to insist that indigenous cultures are backward, stagnant, and showing little vitality. Precisely, the First Congress of Indigenous Peoples held in Pátzcuaro has just shown precisely how political the indigenous question is.

The main reason the Congress was held was undoubtedly President Luis Echeverria's personal interest in indigenous groups, which through the rare alchemy of our political system, has turned into an intensified effort by the National Indianist Institute (INI) to fortify Agrarian Reform programs and to the National Congress of Campesinos (CNC) recent eagerness to incorporate indigenous peoples into its ranks. In 1975, as reported by these two agencies, 60 regional congresses were held to elect the delegations of these Indian groups. How were such congresses held? Who was invited to them? Where were they held and on

[1] Published as one of the chapters in Arizpe, Lourdes (1978): *El Reto del Pluralismo Cultural* (Mexico: Instituto Nacional Indigenista). Unpublished in English. This publisher has disappeared.

L. Arizpe, *Lourdes Arizpe*, SpringerBriefs on Pioneers in Science and Practice 10, DOI: 10.1007/978-3-319-01896-6_11, © The Author(s) 2014

what dates? Mystery, mystery, very PRI-like. Transported in buses, the Popolocas, Pai-Pais, Pimas, Mazahuas and another thirty delegations appeared in Pátzcuaro as if brought there by a magic wand. Again, the secrecy and concealment of the operations foretold a performance to be given by puppets, whose promises of meager concessions would be used again to support this or that pre-candidate and to give luster to this or that agency or organization. Precisely for that reason, a group of anthropologists got ahead of the game and published a protest manifestation in a Mexico City newspaper. But the huge, extraordinary surprise in Pátzcuaro was that this time, the indigenous people took over the right to speak and the management of the congress. This piece of news reached the newspapers with many distortions but the facts speak for themselves.

11.1 Pátzcuaro, Monday Night

Buses continue to arrive, bringing Totonacos from Papantla, Nahuas from Zitlala, Guerrero, and Mayas from Tenabo, Campeche. They were promised lodging: they are put up in enormous circus tents that look very gay, with red and yellow stripes, on the shores of the lake. Would groups of rich farmers or ranchers have been lodged like this? The beds and blankets are to be brought by the army, but they are delayed. They did finally arrive, on Wednesday. In the meantime, the delegations could sleep on 'petates', straw mats, since "they are used to it" said one of the organizers. And the answer of some of the Indians: "Yes, we are used to it." But now others protested forcefully: "They are treating us like cattle." The reason why they all ended up accepting this was made explicit by a Mayan peasant during one of the sessions, "We did come here and we can stand to sleep on the ground and take whatever they give us to eat, but it is because we are hoping to get some benefit from this meeting."

Rumor had it that army trucks had been delayed because they were stopped by the state highway police. That could be, because there was another rumor going around that the governor of the state of Michoacán was vehemently opposed to the Congress. And all that was needed was to read the comments about it in the newspaper *Voz de Michoacán*.

Some delegations are lodged in the dormitory of the CREFAL[2]; others explained "Well, we spent the night better in the truck, just talking." Restrooms and drinking fountains were installed in the nearby areas and three meals a day were served to 2,800 participants in the INI Coordinating Center, one block away. Seated on the grass in the garden or around the tables, the various groups look at each other uncertainly, suspiciously, or with a new feeling of solidarity. Pepe Chambor, chief of the Lacandon, sat talking to Jose Martínez, one of the Popoloca

---

[2] A United Nations institutions for education in Latin America and the Caribbean located in Patzcuaro.

delegates. The women, most of them wearing their traditional costumes, do not talk at all without asking permission from their men. It is very cold and the Tacuates, who live on the Coast of Oaxaca and wear beautifully embroidered short pants, were running to and fro looking for blankets.

Inside the tents, a few ropes indicated territory: Here were the Chatinos, over there the Mazatecos and on the other side, the Amuzgos. At night, there is the din of noisy chatter. Some are playing card games, laying them out on the straw mats. Then somebody begins to play a guitar and a violin and there is dancing, showing off the different dances. But for the chiefs of the delegations and the more politicized young leaders, it is only a time for discussions.

Because there are two types of delegates present: the young and old militants, who talk rapidly, attempting to strengthen their awareness and the ties of political support among the various delegations, "only if we are united and support each other, will we be able to do anything; if not, they will swallow us up again". They walk faster to get together to keep on discussing, absorbing information, and getting organized. The other type of delegate is like this Indian man who had never left his village before. He is always smiling: "Yes, this is very nice, miss, yes, we are very grateful, yes, thank God for having allowed us to come here." "And what do you think about the Congress?" I ask him. "Well, it is very nice, miss, yes we are all like brothers here and we are having a very good time."

11.2 Tuesday Morning

On the dais installed for the inauguration on the island of Janitizio, in the middle of the lake, there are two rows. In the first row, officials take their seats in chairs; behind them, standing in the sun, are the indigenous delegates. The speeches throw out well-known slogans, but they add: "we are here, not to talk, not to tell you what has to be done, but to listen to you, so that now the indigenous people can talk and be heard". Having said this, almost all of the officials get up and disappear, leaving the indigenous leaders to talk alone, with their voices getting lost over the waters of the lake.

The Challenge of Cultural Pluralism (1978)

11.3 Wednesday Morning

During the afternoon of the previous day, the organizational arrangements for the discussion groups had been made quickly by the officials. The discussion leaders had been designated, presentations had been planned, and maybe even the conclusions. But then the organizers of the Congress waited and waited, until finally the Indians showed up. They then announced that they would hold their own sessions and with a "and now please excuse us", they asked all the officials from the various government departments and the CNC, the anthropologists and reporters to leave the room. And then they spoke, for hours and hours. In the end, they announced that only the indigenous people would have the right to speak and to vote in the working groups, which would be led by Indian participants, and that officials, in their status as consultants, would receive the conclusions and petitions from the working groups in due course. Now the delegations that had been wandering around got down to work. The change in the atmosphere is striking; Indian participants hurry around to present their petitions in the working groups; they make the effort to speak in Spanish and to speak in public.

During two days they worked in different sessions. Among those that had the largest attendance were the working groups on Land Ownership, Forestry Issues, Marketing, and Livestock Production. There were frequent comments such as this one by an Opata farmer: "When we protest, when we speak out to say that they are taking our lands away from us, because of that, they want to shoot us. They threaten to kill me every so often." One view in particular kept being repeated: "So that's why I say that there should be a law for everyone and not just for the rich people. Because they are there, with their horses and their pistols, and nobody does anything to them. They kill our people and nobody does anything to them." The same accusations, the same problems: The Lacandon, the Tarahumaras and the Mazatecos and several other groups, to give an example, found that they all had grievances about the same kind of abuse by lumber companies that were exploited the forests in their regions. Peasants who grow coffee found that they face the same obstacles to marketing as their colleagues from other regions. *This was perhaps one of the most important results of the Congress. It confirmed, in the eyes of indigenous people, that their most serious problems are similar or identical because they come from a common source: the fact that they have no legal defense against economic abuse because of their lack of political power.* And it seems to me that it was this awareness, discussed at length during the evenings of the Congress, that began to open up a new horizon for the Indians there. Anyone who has been in contact with indigenous groups could not fail to perceive an unprecedented militancy among the delegates to the Congress. For the first time they insisted on taking the initiative in policies that are applied to their groups and in the leadership of the agencies that carry them out.

11.4 Towards a New Beginning?

Linked to this politicization, a change became very apparent in their perceptions of their history and their cultures. "Now I can say that I am Indian and I don't feel bad about it." The working group on Preservation of the Indigenous Languages and Cultures took on the task of analyzing what the indigenous cultures consist of: Are they distinct traditions, customs, or thoughts? They concluded "that we have our customs, for example, our traditional governments with administrators and judges, or the *teqio* (a type of communal work) and a way to work the land; and these are institutions, just like the mestizo society has its own, like the PRI." Some very old grievances were heard, about how the Spaniards destroyed flourishing peoples and cultures. And a phrase that goes even deeper: "The Spaniards did not conquer us. A woman can be conquered, only by love. The Spaniards invaded us." Can this new idea of Indian cultures gain strength, particularly in a society that still bases its privileges on economic and political oppression of the indigenous people?

The other working groups also held their sessions. In the working group on Indigenous Women, for the first time in their lives Indian women were asked to give their opinion. Very frightened, not quite believing it, with trembling voices, they gathered the courage to speak. One after another, they asked to be respected as women and as indigenous people. "Sisters, we must fight to be respected, so that they will not humiliate us, so that they will not tread on us, as they have always done," said Cirila Sánchez.

Alcoholism was repeatedly mentioned as a common problem for all of them. Beatings and abuse of women and children are everyday occurrences in the communities because of the unrestricted sale of 'aguardientes' and liquor to the men. The women asked for restrictions to be placed on the actions of the producing companies and middlemen, who encourage alcoholism in order to make big profits.

"They should allow women to go to school, because all they want is for them to get married," said a Tzeltal woman from Chiapas. A Tzotzil woman said: "Campesino husbands should be given orientation about the rights of women and learn that they should not insult their wives when they take part in some political, cultural, social and economic decisions." The Triqui group of women requested "Equal conditions for men and women, the right to vote in elections, to inherit property and to manage it according to our own interests." The Mazateca women asked for jobs to be created for women. The requests for roads, postal and telegraph services, social services, hospitals, schools, and handicraft centers for women were repeated many times.

Other working groups dealt with Political Action in the Communities, but this topic got lost in a mystifying political language; Social Security, Education in Indigenous Regions and Youth and Development.

The Plenary Session of the Congress once again was held in secret, and in it the overall conclusions and petitions from the Congress were drafted, and a Charter for Indigenous People was prepared. The petitions included ranged from demands for official recognition—at least at the state level—of indigenous languages, to inclusion of indigenous people in the state legislatures and on the executive boards of organizations that dealt with development programs or to the exploitation of natural resources in indigenous regions.

President Echeverría was present at the closing of the Congress on Friday morning, giving great political weight to its conclusions. And the question now is: has the movement of the indigenous people gained sufficient political drive to enable it to pressure for results or even to exist after President Echeverria leaves the presidency? Or will it just be another attempt that will be carefully erased when he disappears?

The fundamental issue, however, is about the validity of organizing indigenous people politically on the basis of their cultural affiliations. Some people strongly argue that this divides the *campesinos*. Yet, in fact, indigenous peasants have rarely been included in *campesino organizations*. In any case, it is well known that Indians are always left in the last place in the hierarchy of political organizations. Yet, it is a fact that, almost none of the delegates to the Congress would have had the opportunity to have political experience and to share the solidarity of indigenous organizations through the formal political channels. One delegate said it outright: "Do you think that we don't know why they brought us here? We already know all about being hauled here. But, what I say is that, if I stayed there in the village, where not even the buzzards come along, what good is that? It is better for me to have come here and to know what is going on, and now I have something to tell them when I go back." A long time will go by before independent organizations may exist in the Mexican rural regions. Until that happens, the most important effect of the Congress will be the repercussions that it will have among the Indians themselves.

Chapter 12
Perception and Ideology in Interethnic Relations

Whether one considers classificatory schemes as arbitrary or as having a structural significance, they are, in any case cognitively neutral and can be analysed as such by anthropologists.[1] Cognitive categories, though, once they have sprung from observation of reality or from the structures of the mind, can be used ideologically in a specific social context. By ideological I mean that certain traits of objective perception are overemphasized, while others are minimized, thus resulting in a distorted view of reality which can then be used socially for specific ends. In this way, one can say that cognitive categories get caught up in a game of impression management between social groups.

One such category is that of *indio* in Mexico. For example, the *mestizos* in a region I studied insist that the Mazahua people, whom they categorize as *indios*, are lazy, devious, dishonest and alcoholics every single one of them. Is this true? Any anthropologist will immediately jump up and scream that it isn't. And he will waste his time trying to refute a statement which is ideological and not an objective enunciation of reality. At a more abstract level, he will be trying to answer the famous question put forth by Caso (1948) "What is an *indio*?" (¿Qué es un indio?). But if we find that the *indios*, in this case, the Mazahuas, have no more of their share of human weaknesses than the *mestizos* themselves or any other human group, the question we will ask then is 'Why do they believe that?' and thus 'What is the *concept* of *indio*?' Is it only cultural, as our anthropological training leads us to believe, or is it used ideologically, and, if so, why and by whom?

Recent developments in the analysis of the concept of race in Latin America by Pitt-Rivers (1973) and of *indio* by Bonfil(1972); new theoretical propositions by Fredrik Barth (1969) on ethnic frontiers and more scientific explanations of the relationships between indian groups and national societies by both Aguirre Beltran

[1] This text was presented as a paper at the 73rd annual meeting of the American Anthropological Association. It summarizes a more extended paper published only in Spanish "La Ideología del Indio y la Economía Campesina", in: Stavenhagen, Rodolfo, 1976: *Capitalismo y Campesinado en México* (Mexico City: CIS-INAH): 99–129.

L. Arizpe, *Lourdes Arizpe*, SpringerBriefs on Pioneers in Science and Practice 10, DOI: 10.1007/978-3-319-01896-6_12,

(1967) and Stavenhagen (1969, 1970) now allow a clearer understanding of interethnic relations and the ideology that accompanies them.

The long-term fieldwork that I carried out in the Mazahua region of the State of Mexico gave me an unexpected opportunity to analyse this specific problem (Arizpe 1974, 1976). In the 1930s with the Agrarian Reform land was equally distributed to Mazahuas and *mestizos*. Other factors of production were more or lees evenly distributed between them as well. However, forty years later, we find that the Mazahuas are the poorer of the two groups. Is it true, as the *mestizos* and the urban middle classes—whose views predominate in the mass media—contend, that this is due to cultural and psychosocial traits of the Mazahua culture? How has this negative ethnic image affected their economic performance? These are the questions that this paper attempts to answer.

12.1 Theoretical Considerations

Mörner (1969, 1973), in tracing what he calls the 'race mixture' in Latin America, uses the term *indio* throughout as a predominantly racial category. Clearly, for European thought of the sixteenth and seventeenth centuries, a classificatory scheme based on racial considerations was the most convenient and simple to apply. But once the colonial society became rigidly hierarchical with a specific division of labour assigned to each social group, the racially based term of *indio* became a social category. Pitt-Rivers correctly expresses this idea when he says that "the particular concept of race in use in a given context is the expression of a distinction inherent in the social structure, not in the phenotypical distribution of the population" (Pitt-Rivers 1973: 6).

Mörner perceives this change in the meaning of the term but fails to understand it because he confuses its sociological meaning with its ideological use as a racial term. He further confuses the issue by using the logically unsound term of 'socio-racial' category to describe it.

After the wars of Independence the terms used to designate groups resulting from miscegenation such as *cambujo*, *mulato*, *tente en el aire*, etc., generally disappeared (Mörner also points this out but does not explain it). Significantly, though, the term of *indio* became institutionalized.

However, since the racial basis for the term became more and more untenable by the beginning of this century, anthropologists obligingly provided a new basis for it: cultural differences. There is no denying that a cultural component is part of the definition of *indio*, as Pitt-Rivers (1973) shows in an excellent review of the different meanings of the term, but it has become increasingly obvious, as he himself notes, that it is more than just that.

To begin with, Aguirre Beltran, years ago, showed that *indio* is used to designate a caste-like social group rather than a collection of cultural traits: he was the first to point out that a political and economic system of a special nature was superimposed over the 'pure' cultural relationships.

Furthermore, strictly speaking, the term 'Indian cultures' identifies all human groups that are native to America. One must ask then, why should a purely geographical difference be significant from a sociological point of view? Are we then only cultural geographers or are we attempting a sociological analysis? As Stavenhagen perceptively pointed out, from a strictly cultural point of view there are wider differences between the cultures of America than between some of these and Western culture.

In addition to the geographical and cultural components of the term, there is a sociological component which Pitt-Rivers defines as "relationships of a specific type which exist as a function of a total social structure but are not reducible to either culture or class structure" (Pitt-Rivers 1973: 28). Bonfil, on the other hand, considers this to be a political component and suggests that the term 'ethnic group' (*etnia*) be used only when speaking of the cultural heritage of a group, and that the term *indio* be applied exclusively when referring to the political subjection of certain groups to the central political system (Bonfil 1972: 112).

The analytical tools for dealing with this problem at a more theoretical level are provided by Fredrik Barth. He proposes that, by shifting our attention from cultural content to social organization, ethnicity can be freed from being considered a random passing on of bits and pieces of culture and can be treated as a function of the social structure which encompasses several cultural groups. His main point is that ethnic frontiers are maintained and reproduced within the social structure not by processes of acculturation or preservation of the traditional culture, but by inherent conditions within it which make ethnic identity a significant social category.

Using these theoretical tools briefly outlined above, I analysed the data of two Mazahua villages that I studied: San Francisco Dotejiare and Santiago Toxi.

12.2 Ethnicity and the Mazahua

The Mazahua, which number 150,000, live in the northwest corner of the State of Mexico. Together with their Otomi and Tarascan neighbours, with whom they have considerable cultural differences, they are grouped into the category of *indios*. A thorough analysis of the history of the two villages—of which I will only give the conclusions here—showed that their social classification as *indios* has had important effects on their economic situation. These effects have come about through very subtle political, kinship and economic mechanisms.

As with other *indian* groups in Mexico, the Mazahuas have no political representation at any level. Neither the Revolution nor the political organization into municipalities in the 1930s brought them any nearer to participating in political power. Local *mestizos* took over the new political offices of the municipal governments. In the thirties, however, once they felt they had the backing of the central government, the Mazahuas began to take over the political offices in the villages. In some cases this led to bloody feuds, as in the case or San Francisco Dotejiare.

Significantly, as the Mazahuas village rallied round a Mazahua leader, classic examples of *caciquismo* began to emerge. The similarities between the caciques in the two villages I studied—as well as other cases in the region—point to the fact that structural conditions were such that *caciquismo* not only became possible, but, I believe, necessary in the traditional villages. The lack of political representation of the Mazahuas left them powerless to defend themselves against murder and abuse on the part of the *mestizos*. This was made worse due to the corruption of the judiciary system which meant that justice was the parasite of power, i.e. without power there was no possible defence. Countless episodes in the history of the villages bear this out. Thus, the only way to protect the traditional communities was to meet *mestizo* violence with Mazahua violence, i.e. by stepping behind a cacique.

The main point I want to make here is that, although the exploitation and violence against the *indios* has been repeatedly described, the key factor that allows for it is generally glossed over, this being that they are powerless and entirely lacking in political representation because or their classification as *indios*. That is, the mere fact or being classified as such automatically gives impunity to theft, murder and abuse committed against them. Thus, it is easy to see that the Mazahua, consequently will tend to be the poorer: those who can be easily robbed always tend to be. A tradition has been perpetuated then, whereby the ethnicity of the *indio*, or more strictly, his cultural peculiarities are used to deny him any legal rights.

The second important point is that, by virtue of this same classification, the Mazahuas have been excluded from major municipal offices. The *mestizos* then, have monopolized all the benefits of political alliances and economic advantages that are obtained by holding these posts. For example, being named local agent for a bank, or for of the government fertilizer company carries considerable economic advantages. This is negotiated, however, in political circles which exclude the Mazahuas.

There is a third important link between this lack of access to political power and ethnicity, which Barth points out and which is evident in Mazahua villages. In a climate of political hostility and violence, membership in a traditional community becomes a matter of life or death. This results in the adherence to the traditional culture, in this case the Mazahuas culture, not from a cultural obstinacy but as a strategy for survival in a situation where the only protection available is that provided by the Mazahua community and the 'godfather'–*compadre*: the cacique. But the classical ambivalence of *caciquismo*, i.e. its negative aspects, together with the resulting closed character of the communities have retarded the economic advancement of the villages. For, again, economic favours are given out to the small groups of close allies of the cacique; also this same group tends to monopolize the most lucrative activities in the village, i.e. commerce; at the same time production technology and the training in new skills have little diffusion, and so on.

Thus, the political impotence derived from their being considered *indios* has affected the Mazahuas in three ways: it has allowed them to be economically plundered, it has excluded them from lucrative political dealings and posts, as well as from having any leverage in influencing decisions, particularly economic, which could benefit them, and finally, by forcing them to isolate their communities and to support a cacique, it has further excluded them from the economic development of the region.

Another manifestation of this socio-economic exclusion is found within the reading of kinship. Because the Mazahuas are considered *indios* they are conveniently left outside the kinship and *compadrazgo* networks where advantageous political and economic contacts can be made. The Mazahua and the *mestizos* have been, until very recently, strictly endogamous. But with a clear difference: the kin networks of the Mazahuas rarely cross over village boundaries; those of the *mestizos*, in contrast, extend over the main regional villages and towns. Through the latter, the *mestizos* are able to maintain a constant flow of information, personal contacts and alliances all to their benefit. Socially, then, the Mazahua are left out, that is being classified as *indios* constitutes a social stigma which closes the road to social mobility, which in these rural areas usually implies further possibilities for economic mobility.

In regard to values, it is frequently mentioned that certain cultural attitudes of the *indio* are generally responsible for their poverty. Chief among them is the emphasis on ritual and festive expenditures. *Mestizos* in the region often say the Mazahuas spend all their money needlessly on fiestas.

Elsewhere (Arizpe 1974) I showed that this practice in itself was not the case of the less of capital in the village economies. In fact, when Indian villages lived in isolation, the money and goods spent on cargos and fiestas contributed to a faster circulation of goods and gave their economy a dynamism from which a majority of villagers benefited; in a direct manner, goods went to such local specialists as prayer men (*rezanderos*), musicians, dancers, manufacturers of fireworks, flower or wax adornments and so on. But today, these goods flow outside the village. They end up in the hands of (1) the municipal and state governments in the form of taxes (2) the bottled drinks and beer companies (3) the owners of mechanical fairs (4) the *mestizos*, merchants who sell items used in ritual (5) the industrial lists who manufacture consumer goods (6) the Church (7) the orchestras, boxers, magicians and other performers who have made their appearance in village fiestas. All of them take the money collected out of the village, usually to urban centres, which benefit from this inflow of cash. This goes to show that the custom of ritual expenditure in itself is not responsible for this economic drain; it can be attributed to the fact that the village fiestas have been pulled into the orbit of a monetary economy which primarily benefits certain urban sectors.

Another cultural factor cited as causing the poverty of the *indios* is partible inheritance, which, as in the case of the Mazahuas, results in the subdivision of the Land into smaller and smaller plots each generation thus lowering marginal productivity and rendering mechanization and capitalization of agricultural production impossible. As both *mestizos* and the Mazahuas follow this norm of inheritance why then is *minifundismo*—the subdivision of plots and its economic consequences—more prevalent among the latter? I contend that it is because the *mestizo* surplus population has been able to enter economic sectors other than agriculture: specifically they have taken over all commercial service activities; they are occupying newly created jobs such as office-workers, electricians, mechanics and so on in addition to the political offices already mentioned. As a consequence of this the elder *mestizo* sons abandoned agricultural work to enter these jobs and the younger sons have inherited larger plots of land than their Mazahua counterparts.

But the point that must be clearly understood here is that this exclusion of the Mazahuas from economic benefits is not due to some sort of gratuitous social, racial or cultural prejudice. The fundamental factor that gives this ethnic frontier its significance is the scarcity of jobs and economic opportunities within the total social structure. The data from Santiago Toxi, confirm this: a large factory built nearby has provided jobs for the Mazahua young people. As a result, they are enthusiastically giving up their traditional ethnic identity: they have found a place in the occupational structure, which implies social acceptance within the national society. Because these jobs are available in large quantities there is no competition for them between the Mazahua and the *mestizos*, and, as a result, the ethnic frontier is rapidly disappearing in the area.

12.3 Conclusions

The main mechanisms that have brought about differential economic positions for *mestizos* and Mazahuas have been laid bare. In summary, the lack of access to political power on the part of the Mazahuas has left them open to theft and abuse and has excluded them from political circles where considerable economic gains can be made. The social stigma of being *indios* has stopped them from entering into kinship and social networks where favourable political and economic contacts are maintained. Their traditional values in the new setting of a money economy allow a constant drain of capital. By virtue of their identity as *indios* they have also been excluded from the commercial ventures and jobs which have been recently developed and thus, have had no choice but to stick to their ever-shrinking plots of land.

The conclusion of this paper is thus that the Mazahuas economic backwardness is not the result of their cultural attitudes but of the fact that their identity as *indios* has been used to keep them at the end of the queue in a situation of scarce economic opportunities.

The confusion between the cultural and the political (or ideological as I have called it in this paper) content of the term *indio*, has led well-meaning anthropologists to give it a white-wash of neutrality: in fact it is a value-loaded term which is used ideologically as an ethnic frontier in two ways: to keep economic gains within the bounds of the *mestizo* group, and to justify the poverty of the indian groups. Mexican society, caught between the barefoot Mazahua women peddling in the streets of Mexico City and the palatial mansions of certain residential districts must appease its catholic conscience: so it swears that the *indios* are poor because they are *indios*. But the fact that this problem is conveniently ideologized as a cultural phenomenon is no reason for anthropologists to follow this lead and ignore the basic process underlying it: i.e. *that cultural differences become ethnic frontiers in situations of scarce resources*. Which is a way of saying: under this system, somebody must be poor, well then, let it be the *indios*.

References

Aguirre Beltrán, Gonzalo, 1967: *Regiones de Refugio* (Mexico City: Instituto Indigenista Interamericano).

Arizpe, Lourdes, 1974: *Parentesco y Economía en una Sociedad Nahua* (Mexico City: Instituto Nacional Indigenista).

Arizpe, Lourdes, 1976: "La Ideología del Indio y la Economía Campesina", in: Stavenhagen, Rodolfo, *Capitalismo y Campesinado en México* (Mexico City: CIS-INAH): 99–129.

Barth, Fredrik, 1969: *Ethnic Groups and Boundaries* (London: George Allen-Unwin).

Bonfil Batalla, Guillermo, 1972: "El Concepto de Indio en América: Una Categoría de la Situación Colonial", in: *Anales de Antropología*, 9: 35–45.

Caso, Alfonso, 1948: "Definición del Indio y lo Indio", en: *América Indígena*, 8, 4: 239–247.

Mörner, Magnus, 1969: *La Mezcla de Razas en la Historia de América Latina* (Buenos Aires: Paidós).

Mörner, Magnus, 1973: *Estado, Razas y Cambio Social en la Hispanoamérica Colonial* (Mexico City: Sep Setentas).

Pitt-Rivers, Julian, 1973: "Race in Latin America: the Concept of 'Raza'", in: *Archives Européennes de Sociologie*, 14: 3–31.

Stavenhagen, Rodolfo, 1969: *Las Clases Sociales en las Sociedades Agrarias* (Mexico City: Siglo XXI).

Stavenhagen, Rodolfo, 1970: "La Sociedad Plural en América Latina", in: *Diálogos*, 55: 5–10.

Indigenous people in Mexico City – The Case of “Las Marias” (1975)

Chapter 13
Indian Cultures in Mexico City

How do we stop the movement of society for an instant without interrupting it? This, at the heart of it, has always been the basic problem for an anthropologist. How do we capture, describe and analyze the life of a society and at the same time explain how it is changing? Change used to take place gradually and intermittently, easy to keep track of when traditional societies lived in isolation. Today, however, modern societies are increasingly fluid, permeated by and interwoven with Western capitalism. This is why it is more convenient for anthropologists to study a small community by itself, or better yet, an island, like one of the Trobriands: a closed, finite space, without the messiness of cultural limits that blur and bleed into the flow of another culture. Hence, too, the usefulness of a living, for one or two years, but in the end, a finite time, in an Indian community. Because in this slice of time, the life of that society is frozen in the researcher's imagination, and years later she can continue to spin the skein of her interpretation. Cutting out blocks of time and space gives us the advantage of an uninterrupted observation of an encapsulated culture.

Then suddenly, we anthropologists look around us and realize that we have been dragged along, just like all 'others', by this torrent of change. We have lost the comforting structural-functionalist vision: culture as a solid, immovable and balanced edifice. What we see now is a confluence of cultures that mix and are transformed continuously under the influx of others (did we lose our imagined control of time?) Mexican anthropologists suffer from this sensation more intensely because we are both observers and conscious actors in this plural society to which we belong (did we lose our ivory tower?). Hence the interest, the constant urge, from the early times of Mexican social anthropology, to understand this movement, this collective transformation that envelops us and never stops.

We might as well accept it. It falls on us not only to capture, but also to experience, this perpetual flow of culture, like a river that receives tributaries from indigenous traditions, from *mestizo* peasant culture, from the new Mexican urban culture, a flow that empties out and churns in the great showcase that is Mexico City. A city of a thousand cultures, and among them, one that is only faintly perceived: Indian culture.

L. Arizpe, *Lourdes Arizpe*, SpringerBriefs on Pioneers in Science and Practice 10,
DOI: 10.1007/978-3-319-01896-6_13, © The Author(s) 2014

Indian culture? No, that doesn't exist, says the eminent anthropologist Lucy Mair: it is a mirage that confuses anthropologists. All that exists, she would say, are individuals that carry certain beliefs, customs, clothing, and so on. So in the city, we call the Mazahua women who sell fruit in the streets 'Marías'; the Otomí[1] and Mazahua peasants that work as bricklayers 'Josés'. Our eyes and our restless observation take in the proud Tehuana women selling their beautiful golden jewellery, Nahua artisans from Tlaxcala that continue to weave serapes in the lost cities, and men and woman on a corner of San Juan de Letrán street that we come upon speaking an unknown language. But are they really just individuals? Is there no element that standardizes their life in the city, that justifies speaking of a collective? Obviously, there must be something, because there is a generic term to designate it: *indios*. To indicate 'the Other' in relation to Westernized, *mestizo*, urban. Obviously, because people don't normally say the 'Mazahua' fruit vendor, or the 'Otomí' bricklayer, but rather '*indios*' or '*indígenas*' that are, in some vague way, all the same. And it's true: their income levels and lifestyle in this city are similar to the extent that we can accept the existence of this abstraction: the city's 'Indian' cultures.

But even though these exist in Mexico, they have obviously undergone some changes in their passage from rural areas to the city; and obviously, too, they have no precise boundaries. But at what point do these cultures disappear in the city? I would say, at the moment the woman 'servant' says that she does not speak her native language? At the moment when the Otomí bricklayer stops being informal worker and begins to drive a taxi? But the 'servant' still prays to the patron saint of her village, and the taxi driver still chips in for the flowers and fireworks for the village festival. And it doesn't stop there: the urban middle-class man still scrimps and saves and takes on staggering amounts of debt to offer a Rabelaisian feast to his *compadres*,[2] just as any *mayordomo*[3] in an indigenous village would do. And in the upper echelons of power the law of kinship and special bonds continue to govern the distribution of favours, as if it were still an indigenous peasant society. And political speeches are still rife with florid oratory, reverence and cunning disguised as humility, just like those of indigenous village leaders.

Where, then, do Indian cultures in Mexico City begin and end? I should first clarify that the persistence of a few isolated traits does not enable us to speak of an indigenous culture throughout all urban social classes. Strictly speaking, if we take culture to be synonymous with society, it cannot exist independently of an economic infrastructure. But the more popular definition of culture equates it with a system of beliefs and customs. If we take this meaning, it can be said that indigenous culture is lost when its bearers take the step of integrating into the middle class, an entry that requires them to fervently forswear any connection to 'backward customs' and 'Indian things' and to turn their eyes toward all that is

---

[1] Otomí, Mazahua, Nahua and Mixtec are Indian languages spoken in central Mexico.

[2] Close friends, often but not inevitably designating the godparents of one's children.

[3] A village elder, a steward, a ceremonial official.

'white with blue eyes', toward hamburgers and gigantic cars; toward anything, as long as it is foreign.

Curiously, today we are seeing a sort of backlash in the Mexican bourgeoisie: the deliberate adoption of some visible traits of indigenous culture: crafts, clothing –both 'folkloric' as well as authentically indigenous– and dance. But, again, these are merely isolated traits, and they are not accompanied by values, customs or social organization.

Indian cultures in the city are found in their 'purest' form among the lower economic strata (Arizpe 1974). They are usually recent migrants to the city, employed in low-grade jobs and informal labour. The ubiquitousness of these indigenous migrants in the city in recent years is due, first, to a growing 'ruralisation' of migratory groups. In other words, previously it was the upper levels of rural communities that emigrated, but today it is also small-scale indigenous peasants. In the second place, it is due to the fact that while in the past people emigrated alone –generally young women seeking employment as domestics, or young men in search of economic mobility, today entire families are on the move. Obviously, an isolated individual has to acculturate more rapidly in a different medium. A family, in contrast, brings with it the form of life that it left behind, in this case indigenous culture. And, third, ethnic identity among indigenous migrants is being held on to, as they are now having less access to permanent, well-paid jobs, a situation that is linked to the unbridled pace of demographic growth and the decreasing creation of jobs in Mexican industry in the 1960s and 1970s (Arizpe 1978).

Indian cultures in the city keep a cluster of traits such as kinship networks, ritual festivals, work patterns, political views and so on as well as customs: consumption patterns, food preparation, a hierarchy of daily activities, the way home is furnished. Both aspects are associated with a level of income that defines the position of these indigenous people in urban society. To facilitate this explanation and place a more human face on it, I will describe the situation of a typical indigenous family in the city. Beyond analyzing the reasons and characteristics of their socioeconomic position in the city—already described in a previous paper (Arizpe 1972)—I will here attempt to capture the texture of their lives.

The Pineda family lives in the La Merced part of town, in one of its many *vecindades*, old homes of bygone glory where dwellings open out onto a central patio, crisscrossed with clotheslines, having two or three stories of door-lined balconies rising on each side, like a prison. Régulo and Eugenia were lucky to find a room of their own for the family, because their cousins next door all live in one enormous room, with another 20 people who pay to sleep the night, and they all cook together on their little metal grills in the patio. In that room at night, conversations overlap in Mixteco, Mazahua, Nahua and other languages, without anyone paying attention to the others: they are all too concerned about surviving in this sea of people. Almost all of them, except for some women and children, go out at five or six in the morning to sell the products they have brought from their villages, or to buy merchandise at La Merced and take to sell in other markets, or to work as carriers—'diableros'—or labourers.

The Pineda family has lived in the city for ten years, but when you enter the room they occupy, you feel you are entering a house in their village. The only furniture is a table, used not for eating but to hold a few dishes, some prayer candles and ceramic and plush animals. Directly above the table, on a shelf, burns a prayer candle flanked by two vases, one holding some wilted blooms and the other some plastic flowers. Above that hangs a picture of San Santiago, accompanied by a series of religious images superimposed or affixed around him. Below the table is the metal grill, full of charcoal, where the family cooks the same food as in their village: beans, some kind of chilli sauce, occasionally accompanied by some meat, and tortillas. Off to the side is a small gas stove, which broke down some time ago and was never used again.

Despite the meagre conditions in which she lives, Eugenia likes the city. She remembers that as a young woman, in her village, she had to get up at four in the morning to grind the *nixtamal* for tortillas on the *metate*; later on they got a hand mill, but it was still hard work separating the grains of corn, soaking them in lime, grinding the 'masa', patting out the tortillas, lighting the fire with the kindling dry stalks of corn, and heating them. She also had to walk for an hour, bringing one of her daughters with her, to carry water from a stream –that sometimes dried up– to make coffee or tea and watery soup. Now, she is still amazed that she can have water just by opening the faucet in the building's patio, that she has electric lights, and that she can send one of her children down to buy tortillas in the morning. Although she has to put up with police harassment, she still prefers to sell in the street, with her children close by. She enjoys watching people go by and she can give her children a few cents to buy soft drinks and candy. She doesn't like other jobs because she can't stand to be ordered around or told what to do—except by her husband, whom she obeys unhesitatingly.

At night, she gathers with her family and they pass the time chatting in Mazahua about their experiences of the day and, a very important part of their lives, playing with the children. The children dance and tumble on the ground while the adults laugh and applaud. The men are particularly affectionate with their children.

All along the walls there are various straw mattresses rolled up with their blankets. A rope, strung diagonally across the room from one corner of the room to the other, holds almost all of the family's tattered clothing. The rest is kept in cardboard boxes stacked in a corner of the room. At that moment, at two in the afternoon, two children are sleeping on one of the straw mattresses, along with a son-in-law, who just arrived from the village and has not yet found work.

In the other corner lies Eugenia. She had just given birth to her 12th child in the same room. She has never been to the hospital. Because she is at home, she has been asked to care for her two daughters' five children under the age of five. Another two of her sons are less than five years old themselves. Her youngest daughter, recently married, still has no children. Altogether, 17 people live in the room, if you count the newborn child. Eugenia recently sent two other sons, aged 14 and 13, back to her village to help their grandfather with planting the maize crop; another daughter went to work for a rich family in the village, and Eugenia left another two daughters with her mother to help her with the housework and to be raised in the village. She visits them frequently, for example, when she goes back for the village fiesta or for some wedding or baptism, or six months ago, when her mother alerted her through 'the mail'—a person that travels between the village and the city—that she was very ill. Eugenia sent one of her older daughters back to the village on the same day, bringing some money and staying to care for her a few days. But this cost the daughter her job as a guard in a market. Since then, along with her sisters and mother, she sells fruit on Manzanares street. She has been offered work as a domestic, but her husband refused to let her work. He says: "A lot of people criticize us because we let our wives sell in the street, but it would be worse for them to work as servants."

This young man works as a bricklayer's assistant, but he has had little work lately "because now a lot of the *maistros* (informal contractors) bring their own workers from their villages, and they leave us without anything". When he would work for a month or two at a stretch, he could save a certain amount of money and send part to his mother in the village, to make a payment on their land and "to buy clothes for his brothers and sisters". When his father has money troubles, for example, when he drinks too much and uses up the money he had to pay for the animals to pull the plough, he comes to the city to ask his son for money. If his son doesn't have the money, he asks his father-in-law, and pays it back little by little. When the father-in-law needs money urgently, the young man asks a cousin who lives in a *vecindad* two blocks away.

This happened a little while ago, when one of Eugenia's sisters –a widow– died, and her children had no money for her burial. Eugenia spoke with her older son, and told him, "that it was his aunt, and he had to take care of it." So he borrowed money from his brothers-in-law, his father, and an uncle who works in a factory and lives in Ciudad Nezahuacóyotl. Then they found that the 'Rubén Leñero' Hospital was refusing to turn over the body. Eugenia spoke to her husband's aunt, who worked in the city government Department of Markets, and the aunt called a man she knew who worked in the Mexico City government, saying "one of the 'Marías' just died and they don't want to give her to us. And he told me that he would see to it. There in a *vecindad* across from the hospital we went and bought some sheets that a lady had hanging on the line, to wrap the body in". The funeral agency charged them fifteen hundred pesos to take the coffin back to the village, charging extra because they had to traverse a stretch of dirt road connecting the village with the municipal township.

Had it not been for that expense, Régulo would by now have been able to pay off the 800 peso down payment on a little plot of land in the Colonia of Aurora, in Ciudad Nezahuacóyotl. His brother had loaned him money a few months earlier to buy land there, close to where he lives, and offered to organize the construction work, because he is an informal contractor himself. This brother built his own home with the help of brothers, cousins, uncles and nephews who pitched on Saturdays and Sundays. In return he offered them a traditional dinner of *mole* and all the beer and rum they could drink. But now the brother can't loan him any more money because he is over-indebted with some furniture he bought in instalments: a bed, a wardrobe, and a record player, that cost him twelve thousand pesos.

And Régulo had to spend the money his brother loaned him to buy medicine that the pharmacist on the corner prescribed after he suffered a fall on the job, became ill, and was fired. Since then, he suffers from pain in his lungs, and can't work in construction anymore. Once, around eight years ago, he managed to get a job in a yarn factory, even though he didn't know how to read and had neither a grade school diploma nor a vaccination card. "we? Where are we going to get those papers?" he asked. He got the job through the recommendation of a *compadre* from the village. But they paid him very little, and he couldn't stand being locked up eight hours a day in a room, because he was used to working in the fields, with no fixed schedule. Also, around that time he had to go back to the village to help his father with the harvest, and when he came back, the job at the factory was gone.

Since then, he prefers to work as a bricklayer or carrier, because he gets a few days off, as when he lived in the village, and which he can drop for a little while if he needs to go back for a visit. Now he goes out at 5:30 in the morning with his dolly or '*diablo*' to help carry sacks in the market. On good days he can make up to thirty pesos; on a bad day, not even fifteen.

Once when he was working as a bricklayer, they made him go to a government office where, with a great deal of effort, he managed to obtain a Social Security card. But he does not know that this card ("it's put away over there somewhere...") gives him the right to medical services, child care, etcetera—all free. He is nervous about anything related to bureaucracy "because they see that we're poor, they see that we're '*indios*', and they ignore us". Since he and his family have been in the city, most of his contact with government offices has been through the local police. Eugenia has been arrested three times, her two daughters two times each, all for selling in the street. Sometimes they charge a fine of 100 or 150 pesos, money that relatives would lend him. Once they demanded five hundred. They also take away all the fruit they were selling, which was worth another $100. "So I say, how are we hurting them, letting our wives sell in the street? If we do it, it's out of necessity. If there's no work, Miss, what do you want us to do? Steal?" And with that, he takes out his bottle of rum and drinks himself unconscious.

Three days ago a policeman kicked away all the fruit being sold by one of his daughters, and when she protested, he seized her by the hair and dragged her across the street. When I arrived to visit them, she was drunk.

The other government officials they know are market administrators, that either remove them bodily from the market, or charge them two or three pesos a day to allow them to spread their fruit out on a rag on the sidewalk: market inspectors and market collectors work the same way.

For them, the government represents violence, exploitation, and despotic authority against which they are utterly defenceless. And above all, power is so distant from them that they prefer to entrust themselves to closer authorities: the saints and the Virgin of Guadalupe. Every year they take part in the pilgrimage that goes from their region to the Basilica of Guadalupe in Mexico City. But now they no longer ask for rain or a good crop, like they used to back in the village. Now they ask them to keep them free of illness, to protect them from the police, and—though some no longer ask because they have grown tired of asking—to give them a good job.

Is the destruction of their indigenous cultures an inevitable condition of these people's social and economic advancement? Is it actually destroyed, or do just the most visible aspects disappear, while a vision of the world and a series of values persist and permeate Mexican culture? These are fascinating questions, which will require specific research by anthropologists in the future.

At the moment, indigenous culture is flourishing in our city, but in what way? As an emissary of a past that we insist on locking up in books and museums, as a reminder that these people have been left out of the country's economic development? Their strange situation is eloquently expressed in a curious incident: two indigenous women were trying to sell chewing gum at the entrance of the Museum of Anthropology. A uniformed guard hurried out to chase them off, saying:

"Go away. You have nothing to do here." *Eppur* ...

References

Arizpe, Lourdes, 1972: "Rostros Indígenas", in: *Diálogos*, 8,6, 48 (November–December): 15–18.

Arizpe, Lourdes, 1974: "Indian Cultures in Mexico City" in: *Diálogos*, 55 (January–February): 36–40.

Arizpe, Lourdes, 1978: *El Reto del Pluralismo Cultural* (Mexico City: Instituto Nacional Indigenista).

Chapter 14
Zacatipan Kinship Terminology: A Dual Approach

Our main concern in this article is understanding the terminological kinship system of Zacatipan, a nahuat village in the south-eastern region of the Sierra de Puebla.[1] The procedures carried out for this purpose were a componential analysis and a comparative study with other nahuat kinship terminologies.

A componential analysis may function at an ethnographic level, as shown in Goodenough's analysis of Truk and Lapp kinship terminology (Goodenough 1956, 1964) or at a more abstract ethnological level aimed at generalizing. Lounsbury (1964), working at this level in constructing a generative model for Crow-Omaha kinship terminology, described the task of componential analysis as that of specifying the set of primitive elements and the set of rules operating on these, that make up any terminological kinship system. In this study we will be working at an ethnographic level. Therefore, we are concerned only with the isolating of elements in order to represent a particular system and not with finding cross-cultural components and transformation rules. For our purposes we treat the data not as representative of a universal intellectual code, nor as a system of communication, but as a logical system that we can understand better if we can discover the interrelatedness of its elements.

However, we don't divorce it completely from empirical reality. Our study would become a mere exercise in logics if we did—and social anthropology cannot be only a branch of semiology. But the relationship between terminology and behaviour is far from clear. At most, cognitive anthropology cautiously considers some sort of functional relationship between the two levels as axiomatic (for example, Goodenough 1956: 213; Law 1965: 9). Though not proved, it is generally useful to assume that the loss or change of kinterms is directly linked to changes in the social roles.

In these introductory remarks, then, we accept, as expressed by Buchler and Selby, that "these assumptions are method-specific–they do not permit us to make

[1] This chapter was originally published as "Zacatipan Kinship Terminology: A Dual Approach", in: *Canadian Review of Sociology*, 9, 3 (August 1972): 227–237. The permission to republish this article was granted on 25 May 2013 by Prof. Dr. Terry Wotherspoon, Managing Editor, Canadian Review of Sociology.

L. Arizpe, *Lourdes Arizpe*, SpringerBriefs on Pioneers in Science and Practice 10, DOI: 10.1007/978-3-319-01896-6_14, © The Author(s) 2014

general statements about the nature (in its entirety) of the phenomena under observation" (1968: 45). "Componential analysis thus doesn't require nor prove the extensionist theory" (Buchler/Selby 1968: 45). We use it only as a heuristic procedure.

In the second part of this article we include the results of comparing Zacatipan kinship terminology with those of linguistically similar systems—though not necessarily structurally similar. Several contemporary nahuat and nahuatl systems were taken into account but we especially compared results with Joan Law's excellent ethnolinguistic study on nahua kinship terminologies.

14.1 Componential Analysis

The principles expressed by Lounsbury (1956, 1964) and Goodenough (1956, 1964), and their application to several kinship terminologies (Goodenough 1964; Lounsbury 1964; Law 1965), will serve as a basis for our analysis.

Petrona Xalaticpac, the countess of Zacatipan, Lourdes Arizpe's best informant in 1969

The objective of a componential analysis is, through an inductive process, to isolate the minimum number of elements of a system that will describe it sufficiently. We assume that there is a system which can be represented in a series of contrastive sets. As Goodenough describes it: "We presume that all the specific relationships that a term may denote constitute a set (its designatum) such that there are discrimination criteria which differentiate the set as a unit from the sets of possible denotata of other kinship terms" (Goodenough 1964: 223). The components we shall look for, then, are not the minimal units of the system but the criteria used to differentiate between them. It is vital to understand that a kintype "does not denote a person or a kind of person, but a conceivable relationship between persons" (Goodenough 1964: 222).

We begin by defining the universe of kinship in Zacatipan terminology. The term *queniquitta*[2] designates all individuals who trace a consanguineal, affinal or adoptive link to ego. It is highly significant that its literal translation is 'how I see him'. That is, ego is the explicit starting point of the relationship. It implies that as long as ego 'sees' an individual as having a kinship tie with him, he belongs to his universe of kinship. Furthermore, the centre-point of this universe shifts according to who the speaker is, since it can be conjugated in the first, second—*quentiquitta*—and third—*queniquitta*—person singular. An adequate graphic representation is: "Each person sees himself spiderlike at the centre of his own web of kinship, but it is a web occupied by spiders like himself, each looking outwards at the others" (Barnes 1967: 102; Epstein 1967; Brockway 1969).

The universe of *queniquitta* is partitioned into two broad categories: *ichancahuan*,[3] and *nochi chancayetoni.*[4] But here we run head on into a problem: the criterion for differentiating between the two is residence, not kinship. Literally, *ichancahuan* means 'the people in his house' and designates collateral relatives living in other households. *Nochi chancayetoni* means 'all people (who) are in the house', that is, co-residents. It applies then to the domestic group, whether the kinship link between its members is consanguineal or affinal.

No terms exist for 'family' as an elementary unit linked by blood ties or for 'lineage' as an extended group of consanguineal relatives. This is certainly connected with the fact that the economically and socially operational group is the domestic unit. The kindred is not discrete nor corporate; it does not carry out important functions. The household, on the other hand, presents a united front vis-à-vis society. Domestic groups act corporately, pooling land and labour resources; income is handled in one purse and, significantly, it is unthinkable for a household to have more than one hearth. "El que quiera comer aparte, que viva aparte" they say. ("S/he who wants to eat separately should live separately").

---

2 From *quen*—abbreviated form of *quemej:* 'how'; *-(n)iquitta,* 1st person singular present tense verb *itta:* 'to see', with *–qui–,* 3rd person singular pronoun of object.

3 From *i-* 3rd *person singular possessive prefix;—chan—:* 'house'; *—cahuan—*: agentive plural.

4 From *nochi:* 'all'; *chan-:* 'house'; -ca, locative suffix: 'in'; and *-yetoni,* agentive form of the verb *yetoc:* 'to be'.

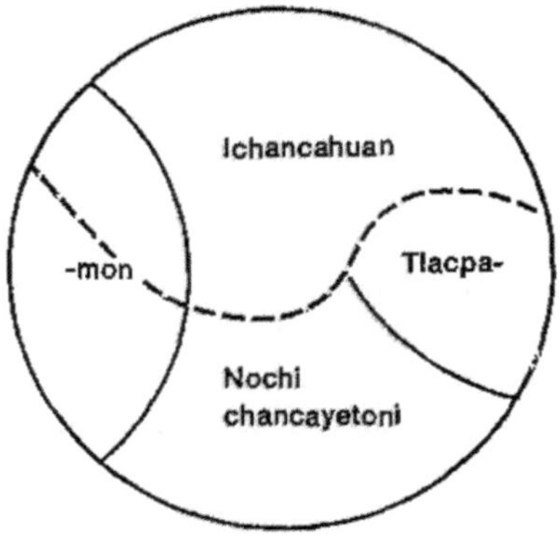

Fig. 14.1 Queniquitta

The term for adoptive kinship is *tlacpa-* added to the kinterm describing the relationship of the person to ego. For example, *tlacpatat* ('tat': father) means stepfather.

Affinal positions in relation to ego are designated by the root-term *-mon* attached as a prefix or suffix to the appropriate nuclear kinterm.

Based on the above data the design shown in Fig. 14.1 was constructed. It tries to represent as closely as possible the concept of the people of Zacatipan have of kinship. The dotted line separating the two main divisions is thus highly mobile.

Lourdes Arizpe's young godchildren in Zacatipan in 1969

Table 14.1 Mapping of the kinship terms

Lexemes[1]	Denotata[2]	English equivalent
Pitox[3]	FFF, FFM, MFF, MFM. . .	Great-grandparent
	SSS, SSD, SDS, SDD. . .	Great-grandchild
Hueitat	FF, MF	Grandfather
Hueinan	FM, MM	Grandmother
Totat	FFB, FMB, MFB, MMB	Great-uncle
Tonan	FFZ, FMZ, MFZ, MMZ	Great-aunt
Tat	F	Father
Nan	M	Mother
Taytzin	FB, MB, FZH, MZH	Uncle
Ahui	FZ, MZ, FBW, MBW	Aunt
Icniu	B, Z	Sibling
(hueltiu)	Z	Sister
Primoicniu[4]	FBS, FBD, MBS, MBD FZS, FZD. . .	Cousin
Sobrino[5]	BS, BD, ZS, ZD	Nephew-niece
Pili[5]	S, D	Child
Conet	S, D	Child
Oquichpil	S	Son
Suapil	D	Daughter
Tayecanque	1st S	First son
Teojca	2nd S	Second son
Teexca	3rd S	Third son
Xocoyot	Youngest S	Youngest son
Ixhuiu	SS, SD, DS, DD	Grandchild
Tagat	H	Husband
Cihuat	W	Wife
Montat	HF, WF	Father-in-law
Monan	HM, WM	Mother-in-law
Huejpol	HB, WB	Brother-in-law
Hues	HZ, WZ	Sister-in-law
Mompoi	HBW, HZH, WBW, WZH	No equivalent
Cihuamon	SW	Daughter-in-law
Monticau	DH	Son-in-law

[1] All terms are in singular; possessive prefixes, no-(tat), mo-, i-, etc., and reverential suffixes, (ta')-tzin, etc., have been omitted

[2] Designators are as follows: F-father; M-mother; B-brother; Z-sister; S-son; D-daughter; H-husband; W-wife

[3] Following Mexican general use, we give the alphabetical signs their value in Spanish. The main differences from English are as follows: 'x' pronounced as 'sh'; 'hu' as 'w'; 'qu' as 'k'; 'j' as aspired 'h'; 'z' as 's'

[4] First particle: primo, borrowed from Spanish

[5] Term borrowed from Spanish

The kinship terms have been mapped in, Table 14.1 as lexemes, taking the definition of the latter as "any utterance whose signification does not follow from the signification and arrangement of its parts" (Goodenough 1956: 199). Several of them are compound lexemes, i.e., *hueitat* (‘*huei*’: big; ‘*tat*’: father): literally ‘big father’, but their meaning in any case depends on their position as elements in a set and not on their lexemic composition. In addition, several lexemes have been borrowed from Spanish, a fact that points out significant problems.

All possible denotata of each kinterm are listed, using conventional single letter designators, and the English translation label is also provided.

Out of the range of possible denotata we have extracted a focal kintype for each term. Then we proceeded to discover its significatum, that is, the criteria that differentiate the class designated by the kintype from all other classes. These components—or abstract principles, as Conklin calls them—we symbolize as follows:

1. Generation: G. (a) Same as ego —G0; (b) ascending —G + 1, G + 2, G + 3; (c) descending —G−1, G−2, G−3.
2. Lineality: c. (a) Direct lineality—C0; (b) first degree collaterally—Cl; (c) second degree collaterally—C2.
3. Sex: (a) Unspecified—S0; (b) male—S1; (c) female—S2.
4. Relative age: R. (a) Eldest—R1; (b) second—R2; (c) third—R3; (d) youngest—R4.
5. Affinity, (a) Direct—Al; (b) one mediator—A2; (c) two mediators—A3.

With the use of these symbols the data can now be most effectively and sufficiently arranged into componential sets. These are then contrasted in a series of paradigms. A paradigmatic classification arranges units which are known (1) to share a certain common feature, and (2) to constitute a contrast set (Conklin 1964: 40). This method follows Goodenough’s concept that “any set of linguistic forms, whatever their shape, which signify complementary sememes may be said to belong to the same paradigm” (Goodenough 1956: 197).

Pitox	G + 3, S0, C0	Primoicniu	G0, S0, C1
Hueitat	G + 2, S1, C0	Pili, conet	G−1, S0, C0
Hueinan	G + 2, S2, C0	Oquichpil	G−1, S1, C0
Totat	G + 2, S1, C1	Suapil	G−1, S2, C0
Tonan	G + 2, S2, C1	Tayecanque	G−1, S1, C0, R1
Tat	G + 1, S1, C0	Teojca	G−1, S1, C0, R2
Nan	G + 1, S2, C0	Teexca	G−1, S1, C0, R3
Taytzin	G + 1, S1, C1	Xocoyot	G−1, S1, C0, R4
Ahui	G + 1, S2, C1	Sobrino	G−1, S0, C1
Icniu	G0, S0, C0	Ixhuiu	G−2, S0, C0
(hueltiu)	G0, S2, C0	Pitox	G−3, S0, C0

It is evident that this procedure has allowed us to organize our material in the most parsimonious and systematic manner. The nature of the interrelationships between the kintypes will emerge even-more clearly by arranging the sets into paradigms that will point out complementary and contrasting categories.

Paradigm II clearly indicates that terms not specifying sex are mutually exclusive with those expressing sex differentiation. The one exception to these complementary sets is G−1 with three kintypes. We can suggest the following explanation: the kinterm G−1, S0, C0 is applied generally to any little boy or girl in the community. This points to a *social* use of the term not relating exclusively to kinship. This is connected to a linguistic feature of nahuat.

Male	Unspecified	Female
G + 2, S1, C0	G−3, S0, C0	G + 2, S2, C0
G + 2, S1, C1		G + 2, S2, C1
G + 1, S1, C0		G + 1, S2, C0
G + 1, S1, C1		G + 1, S2, C1
	G0, S0, C0	(G0, S2, C0)
G−1, S1, C0	G0, S0, C1	
	G−1, S0, C0	G−1, S2, C0
	G−2, S0, C0	
	G−3, S0, C0	

Extensive use of possessive pronouns in the form of prefixes frequently changes the meaning of the word. For example, *pili* generally designates any boy or girl; but if the first person singular possessive prefix is attached to it, i.e., *nopil,* it becomes 'my son'.

Paradigm III: Lineality and collaterally

Lineal	Collateral
G + 3, C0	
G + 2, C0	G + 2, C1
G + 1, C0	G + 1, C1
G0, C0	G0, C1
G−1, C0	G−1, C1
G−2, C0	
G−3, C0	

The asymmetry in the paradigm points to the introduction of kintype G + 2, C1 as discrepant. We can understand it only in the cultural context. The same attitude of respect that is due to grandparents is expected for the grandparents' siblings. Although their descendants are no longer considered relatives of ego, grandparents' siblings are treated as close relatives since they are the uncles and aunts of one's father or mother. In general, behaviour towards siblings of grandparents follows the rule of honouring and respecting all elderly people, which is characteristic of nahua culture generally.

Paradigm III also provides an interesting example of how historical explanation can complement a formal account in understanding a kinship system. Structurally the contrast at the level of G0 and G−1 is symmetrical, but we lose in understanding if we ignore the fact that both collateral kin- types of these generations are borrowed from the Spanish language. If we assume that they are recent innovations, we can suggest a tentative hypothesis: that Zacatipan kinship terminology did not differentiate previously between lineal and collateral relatives.

Further evidence supporting this hypothesis will be discussed later.

Paradigm IV: Affinity

G0, S1, A1	G−1, S1, A2	G0, S0, A3
G0, S1, A1	G−1, S2, A2	
	G0, S1, A2	
	G0, S2, A2	
	G−1, S1, A2	
	G−1, S2, A2	

Affinal kintypes are uniformly contrastive following the criteria of generation and differentiation of sex, as well as link through one or two mediators. The notorious exception which poses somewhat of a problem is the kinterm *mompoi.* It does not specify sex and is reckoned through two marriages and a sibling link. It is the only instance of a consanguineal link being joined to two affinal ties. We can understand it better in the ethnographic setting. Postmarital patrilocal residence makes it normal for several married brothers to live together. Their spouses are *mompoime;* they have close contact and co-operate for household and agricultural chores. In feast-giving not only the co-resident brothers' spouses share the work, but those of brothers living independently are expected to help as well. It is also frequent for husbands of sisters, although living apart, to join in a reciprocal labour group. In fact, it is said that men prefer to work with male affines instead of with their own brothers because conflicts over inheritance rights or authority are less likely to arise. However, the fact that such a social role exists and is important in the activities of the domestic groups does not prove that there is a causal link between role and kintype. For the time being we can only point to the apparent relationship between the two.

Thus, a componential analysis has given us an understanding of the internal consistency of the terminological kinship system and has provided valuable guidelines to essential features of the organization of domestic groups. We now want to complement our findings by comparing the kinship system of Zacatipan to other nahua kinship terminologies. Our purpose here is more cultural and historical in emphasis.

Honoured by 'mayordomo' to carry the Saint during a procession in Zacatipan in 1969

14.2 Comparative Analysis

The basis for our comparative analysis was provided by data from two nahuat villages belonging to the same geographical region: Zacapoaxtla (Key 1953) and Nauzontla (McQuown 1954). Another two villages belong to the main linguistic variant, nahuatl: Atla (Montoya 1964) lies in the western part of the Sierra of Puebla, and Tlaxcalancingo (Olivera 1967) in the southern part of the State of Puebla. But most of our comparative material we took from Joan Law's excellent collection of nahua kinship terminologies and from her penetrating examination of the social organization of classical Mexican society and of a modem village, Mecayapan. All these studies provided an excellent background against which to put Zacatipan kinship organization in relief.

The three kinship systems reported in the Sierra of Puebla are structurally similar. The four—with Zacatipan—belong to the Hawaiian type: for Nauzontla and Zacapoaxtla we lack data on residence, but those of Zacatipan and Atla are associated to the Patri-Hawaiian subtype and are evolving towards Neo-Hawaiian. The characteristics and consequences of this type of kinship organization can be found in Murdock (1949: 228).

In comparing kinterms individually we found in the other three systems a lexeme with sex distinction for sister: *hueltiu*, or a lexemic variant. It is interesting that in Zacatipan it exists only as part of a set expression: *nech- maca mohueltiu*: 'give me your sister' which is used jokingly or offensively—depending on the tone—between men. Thus, although the term exists, it no longer functions for kinship for it has acquired a definite offensive undertone. This is the reason I have included it in parenthesis in Table 14.1.

In the other systems there is also a distinction of sex of the speaker, which alters the terms used to address a brother and a sister. But we found no evidence, either linguistic or in the domestic organization in Zacatipan, that it existed previously. Any attempt at explaining it would therefore spring entirely on an improvable assumption: i.e., that this kintype existed previously in the terminological system.

Comparing the structural principles of the social organization proved much more rewarding. Joan Law in her study successfully correlated distinctive features of the Classical and the Mecayapan kinship terminologies to general principles of their social structure. She gives the following components for their kinship systems, which I have also traced in Zacatipan (Classical and Mecayapan; Law 1965: 125).

Classical. (1) Extension of lineal types to four generations with reciprocity in the fourth. (2) Age-grading of offspring. (3) Distinction between col-lateral kin of Cl and C2 degrees with G−1 limits.

Mecayapan. (1) Extension of lineal types to only two generations. (2) Absolute ranking by age of collateral kin overriding generation. (3) Inclusion of collateral kin of C1 and C2 with G−l limits in one kinclass.

Zacatipan. (1) Extension of lineal types to three generations with reciprocity in the third. (2) Age-grading of offspring. (3) Distinction of collateral kin of C1 and C2 degrees with G−1 limits.

The first system is classified as Eskimo, while Mecayapan and Zacatipan belong to the Hawaiian type. At a formal level, the transition from Eskimo to Hawaiian results from the adoption of distinctive terms for collateral kin of ego's generation. We pointed this out in Paradigm III, suggesting that the terms for cousin, and nephew and niece, were introduced in modern times. This hypothesis rests on linguistic evidence on the one hand, and on ethnographic fact on the other. It is insufficient to state that this was due to the influence of Spanish kinship terminology, in which cousin 'primo' and nephew 'sobrino' are distinctive kintypes. Diebold, quoted by Law (1965: 111) considers it a historical problem of whether the role itself resulted from cultural contact. We cannot agree because one is then assuming that the role is an independent variable; that as a need arose, and a Spanish example was provided, the role of cousin was plugged into the network of social positions. This again breaks the kinship system into independent elements. Instead we propose that the change occurred inside the social system, due to structural modifications in the development of the domestic group.

Older informants in Zacatipan were unanimous in lamenting that the family was breaking up. But its breakdown is perhaps not as recent or as drastic as it seems to them. Years back, they say, sons and their families lived in their father's house until he died or until their own sons married. At that time records show that land was plentiful, plots were not legally partitioned, and thus there was no friction between siblings over inheritance of land. There was no market for labour so that all the income of the household came from the land, to the cultivation of which all members of the domestic group contributed. Neither was there a broad market of goods, so that sons were not inclined to ask for their share of the household income in cash. The isolation and homogeneity of the community also contributed to minimize the influence of economic incentives from the outside world.

In the past fifty years, the boom in coffee production, and, partly as a consequence of it, the massive migration of *mestizos* that has swept the region, abruptly landed Zacatipan in the national economy. Consequently, land has become scarce; the rule of inheritance in equal shares still functions, so that plots are becoming smaller and smaller. As sons despair over the small holdings they will receive they turn to wage labour, the income from which they no longer wish to pool in the household finances. As a new range of goods becomes available, and as new values stressing economic accumulation and the possession of goods as status symbols gain acceptance, young men are eager to farm independently and earn their own income as quickly as possible.

Now, then, extended patrilocal residence results in sons growing up in the same household as cousins, where they are all addressed as ‘child’. Since married brothers lived together, then ego’s brother’s children would also be designated by the term for child. For ego, therefore, there would be no difference between his brothers and his cousins. The same term, then, *icniu,* could be applied to both.

But if, due to the historical circumstances we have mentioned, young couples now tend to live neolocally, their children, growing up in separate households, would not consider themselves brothers. Nor will an individual regard his brother’s children as his own.

The distinction between cousins and brothers, then, could be introduced *either* by developing a nahua lexeme or by borrowing the Spanish one. An example of the first type is the Tlaxcalancingo terminology, where cousin is designated by *huejcaicniu* (*huejca*: ‘far’; *icniu*: ‘brother’; Olivera 1967: diagram). An example of the second is Zacatipan, where the Spanish term ‘primo’ has been added as a prefix to the nahuat kinterm for brother: *primoicniu.*

Thus the role is not the result of cultural contact but of internal structural change. The fact that native kinterms were also evolved for the role also indicates that it was not due solely to the influence of the Spanish language.

Law relates the structural principles of the terminological system to features in the social organization. The lineal reckoning of kinship to four generations, and the age-grading of offspring in the Class terminology, she sees linked to marked social stratification. An individual’s position depended on who he was descended from and on his order of birth, in a system where succession and inheritance passed down the agnatic line.

In contrast, in peasant classless Mecayapan succession and inheritance are not rigidly prescribed. This is due to communal ownership of land and, she suggests, to the breakdown of an earlier clan or barrio organization. This loss of clan or lineage organization, associated with classical times, would have brought about the disappearance of the bifurcating type system and the emphasis on lineal types and age-grading.

However, isn’t it too broad a jump from a linguistic category to a macro-sociological principle of organization? The two phenomena can be said to belong to different orders of magnitude. And the assumption is made that specific behaviour consistently follows the linguistic label. Zacatipan is a peasant, classless society as well. Yet, in the kinship terminology, we find age-grading of offspring.

We cannot understand it in relation to the major phenomena of social stratification. The explanation lies, rather in the composition of the immediate domestic group.

Ideally, inheritance is equally distributed between siblings. Thus, relative age is overtly unimportant in the acquiring of inheritance rights. In practice, though, brothers keep the largest share, and all the inheritance if sisters reside virilocally. Moreover, it is the eldest son who succeeds, and so on. Ultimogeniture exists in that the youngest son is expected to remain in his father's house, but he is always subject to his eldest brothers' authority. Significantly, the brother holding the position of head of the household has a strong hand in deciding when and how to partition the paternal holdings. Succession, then, does make a difference in the pattern of distribution of inheritance. Hence the important function of age-grading even in a classless society. It is necessary for regulating internal domestic succession, though not for sequential accession to power between groups in a social hierarchy.

This brings us to an essential point: there is no reason, then, why age- grading in Classic kinship terminology should be associated with a lineage or clan organization.

14.3 Conclusions

In sum, Zacatipan kinship terminology is ego-centred and belongs to the Hawaiian type. The kinship universe is delimited by consanguineal, affinal, and adoptive ties. The abstract principles that differentiate the significant of the kintypes are generation, sex, collaterality, affinity, and relative age. But we had a problem in that residence must also be considered a componential category.

The linguistic origin of the kinterms for collaterals of ego's generation and first descending generation indicated that they were introduced in modem times. We presented linguistic and ethnographic evidence showing that their adoption was due to internal structural change in the developmental cycle of the domestic group resulting from recent changes in social and economic conditions. Only to a lesser extent was it due to the influence of Spanish kinship terminology.

The correlations between principles of kinship terminology and features of social structure given by Joan Law for Classic Mexican society and modern Mecayapan were compared to our data from Zacatipan. We suggested that the order of magnitude between terminological category and macro-sociological principle is too discrepant to allow a direct correlation to be made between the two. Instead, we proposed that age-grading is correlated to the pattern of succession and inheritance inside the domestic group rather than to major social stratification.

This type of comparative structural study has further interest for historical reconstruction. The data we analysed on kinship terminology and development of the domestic group may allow us to predict that, owing to certain structural conditions, residence in Classic nahua society was patrilocal. And we have also seen that age-grading need not be directly correlated with a clan organization.

References

Barnes, John Arundel, 1967: "Genealogies", in: Epstein, Arnold Leonard (Ed.): The Craft of Social Anthropology (London: Tavistock): 101–128.
Brockway, Earl, 1969: *Términos de parentesco del náhuatl. Dialecto del Norte de Puebla* (Mexico: Anales del INAH).
Buchler, Ira R, Selby, Henry A., 1968: *Kinship and Social Organisation: An Introduction to Theory and Method* (New York: Macmillan).
Conklin, Harold C, 1964: "Ethnological method", in: Goodenough, Ward H. (Ed.): Explorations in Cultural Anthropology: Essays in Honor of George Peter Murdock (New York: McGraw Hill): 25–55.
Epstein, Arnold Leonard, 1967: *The Craft of Social Anthropology* (London: Social Science Paperbacks).
Goodenough, Ward, 1956: "Componential analysis and the study of meaning", in: *Language*, 32, 1: 195–215.
Goodenough, Ward, 1964: "Componential analysis of Konkama Lapp kinship terminology", in: Goodenough, Ward (Ed.): *Explorations in Cultural Anthropology: Essays in Honour of George Peter Murdock* (New York: McGraw-Hill): 221–238.
Key, Harold; Key, Mary, 1953: *Vocabulario mejicano* (Mexico City: Instituto Lingüístico de Verano).
Law, Joan, 1965: "An Ethnolinguistic Study of Nahua Consanguineal Kinship Terminology" (MA dissertation, University of Texas).
Lounsbury, Floyd, 1956: "A semantic analysis of Pawnee Kinship usage", in: *Language*, 32, 1: 158–194.
Lounsbury, Floyd, 1964: "The formal analysis of crow- and Omaha-type kinship terminologies", in: Goodenough, Ward (Ed.): *Explorations in Cultural Anthropology: Essays in honour of George Peter Murdock* (New York: McGraw-Hill): 351–393.
McQuown, Norman, 1954: "La Fonémica de un dialecto Olmeca-mexicano de la Sierra Norte de Puebla", in: *México Antiguo*, 6: 61–72.
Montoya, Jose, 1964: *Atla: Etnografia de un Pueblo Nahuatl* (Mexico City: INAH).
Murdock, George Peter, 1949: *Social Structure* (New York: Macmillan).
Olivera, Mercedes de V., 1967: *Tlaxcalancingo* (Mexico City: INAH).

Universidad Nacional Autonoma de Mexico (UNAM)

The National Autonomous University of Mexico (UNAM) was founded on 21 September 1551 under the name 'Royal and Pontifical University of Mexico'. It is the biggest and most important university in Mexico and in Ibero-America.

As part of the *Centennial Celebrations of Mexican Independence*, the National University was officially created on 22 September 1919. With the intention of widening educational opportunities in the country, the effort to launch the National University, though often hampered by adversity, was initially spearheaded by Congressman Justo Sierra in 1881. His vision finally materialized in 1910 with the inauguration of the *National University of Mexico* at a ceremony held in the National Preparatory School Amphitheatre, where as Secretary of Public Instruction, Sierra told the audience that the thrust of the National University's educational project was to concentrate, systematize and disseminate knowledge in order to prepare the Mexican people for the future. One hundred years after the creation of the University, Justo Sierra's inaugural address still rings true: *(…) we are telling the university community today that truth is unfolding: go seek it (…) you have been charged with the realization of a political and social ideal which can be summed up thus: democracy and liberty.*

The fundamental aim of UNAM is to serve both the country and humanity, to train professionals, to organize and carry out research, mainly on national problems and conditions, and to offer cultural benefits in the broadest sense possible.

Website: www.unam.mx.

L. Arizpe, *Lourdes Arizpe*, SpringerBriefs on Pioneers in Science and Practice 10,
DOI: 10.1007/978-3-319-01896-6,

Centro Regional de Investigaciones Multidisciplinarias (CRIM)

The Regional Multidisciplinary Research Center (CRIM) is an academic institution ascribed to the Coordination of Humanities at the National Autonomous University of Mexico (UNAM). It is located in the City of Cuernavaca on the Morelos Campus of UNAM. Its objectives are:

1. Focus on multidisciplinary research in social sciences, humanities and other disciplines, mostly aimed at tackling specific problems at the local, regional, national and international levels, and their implications within globalization processes.
2. Contribute to the creation of knowledge in relevant and innovative arenas addressing social problems that require the convergence of different disciplines for their study.
3. Contribute to the development of a multidisciplinary approach to humanities, and focus on the development of innovative theoretical and methodological perspectives.
4. Participate in educational programs so as to contribute to the academic training of professionals in social sciences, humanities and other disciplines.
5. Disseminate by all possible means the results of CRIM's research projects.

Website: http://www.crim.unam.mx.

L. Arizpe, *Lourdes Arizpe*, SpringerBriefs on Pioneers in Science and Practice 10,
DOI: 10.1007/978-3-319-01896-6, © The Author(s) 2014

About the Author

Lourdes Arizpe is a professor at the Regional Multidisciplinary Research Center of the National University of Mexico. She received an MA from the National School of History and Anthropology in Mexico in 1970 and a PhD in Anthropology from the London School of Economics and Political Science, UK, in 1975. She has pioneered anthropological studies on migration, gender, rural development, and global change and culture in Mexico, in Latin America, and in international research groups, both academic and policy-oriented. Professor Arizpe taught at Rutgers University through a Fulbright grant in 1979 and carried out research in India and Senegal with a John S. Guggenheim grant in 1981. She was director of the National Museum of Popular Cultures in Mexico 1985–1988. She was elected President of the National Association of Ethnologists of Mexico in 1986 and Secretary to the Mexican Science Academy in 1992. Professor Arizpe was Director of the Institute of Anthropological Studies at the National University of Mexico, was elected President of the International Union of Anthropological and Ethnological Sciences in 1988, and successfully organized its World Congress in Mexico in 1993.

Lourdes Arizpe became a member of the United Nations Commission on Culture and Development, and soon afterwards was designated Assistant Director General for Culture at UNESCO 1994–98. She was elected President of the International Social Science Council for 2004–2008, and participated as a member of the Academic Faculty of the Global Economic Forum at Davos, Switzerland 2000–2004. At the United Nations Institute for Research on Social Development, she was Chair of the Board 2005–2011 and a member of the Committee for Development Policy of the Economic and Social Council. She is a member of the Board of Trustees of the Library of Alexandria in Egypt.

L. Arizpe, *Lourdes Arizpe*, SpringerBriefs on Pioneers in Science and Practice 10,
DOI: 10.1007/978-3-319-01896-6, © The Author(s) 2014

Lourdes Arizpe became an Honorary Member of the Royal Anthropological Institute of the UK in 1995, and has received the Order of "Palmes Académiques" from France in 2007, the Award for Academic Merit of the Universidad Veracruzana in Mexico, and an Honorary Doctorate from the University of Florida at Gainesville in 2010.

About the Book

This book presents major texts by Prof. Dr. Lourdes Arizpe Schlosser as a *Mexican Pioneer in Anthropology* on the occasion of her seventeenth birthday. She is a leading researcher on indigenous people, an innovator in women's studies, a global scientific leader who has inspired the international research and policy communities. Throughout her distinguished career she has analysed ethnicism and indigenous peoples, women in migratory flows, cultural and social sustainability and intangible cultural heritage as social capital, placing these issues on the world agenda for knowledge and policy. Several of the 12 major texts in this volume have been published since 1972 in the US, Europe, Latin America and India; some were first published in Spanish and are available in English for the first time. This anthology also includes recent unpublished texts on culture, development and international cultural policy delivered at high-level international meetings.

L. Arizpe, *Lourdes Arizpe*, SpringerBriefs on Pioneers in Science and Practice 10,
DOI: 10.1007/978-3-319-01896-6, © The Author(s) 2014

MIX
Papier aus verantwortungsvollen Quellen
Paper from responsible sources
FSC® C105338

If you have any concerns about our products,
you can contact us on
ProductSafety@springernature.com

In case Publisher is established outside the EU,
the EU authorized representative is:
Springer Nature Customer Service Center GmbH
Europaplatz 3, 69115 Heidelberg, Germany

Printed by Libri Plureos GmbH
in Hamburg, Germany